OCR

Physics

for A2

Chris Mee

Mike Crundell

Brian Arnold

Wendy Brown

D1439323

HODDER
EDUCATION
AN HACHETTE UK COMPANY

personal tutor

Hachette Livre UK's policy is to use papers that are natural, renewable and recyclable products and made from wood grown in sustainable forests. The logging and manufacturing processes are expected to conform to the environmental regulations of the country of origin.

Orders: please contact Bookpoint Ltd, 130 Milton Park, Abingdon, Oxon OX14 4SB. Telephone: (44) 01235 827720. Fax: (44) 01235 400454. Lines are open 9.00–5.00, Monday to Saturday, with a 24-hour message answering service. Visit our website at www.hoddereducation.co.uk.

© Chris Mee, Mike Crundell, Brian Arnold, Wendy Brown 2009
First published in 2009 by
Hodder Education, an Hachette UK Company,
338 Euston Road
London NW1 3BH

Impression number   5 4 3 2 1
Year   2012 2011 2010 2009

Cover photo © Tyler Boley/Workbook Stock/Jupiter Images
Typeset in 11/13pt Times Ten Roman by Macmillan Publishing Solutions
Printed in Italy

A catalogue record for this title is available from the British Library

ISBN 978 0 340 967812

Contents

Preface

This book is a revision of the A2 material in *AS/A2 Physics*, amended and updated so that it is suitable for students following the OCR Physics Syllabus A. It is intended to follow directly from *OCR Physics for AS*.

All the assessable learning outcomes are covered in this book, with some re-ordering of topics in the interest of a logical teaching order. In a few cases, the material of the book goes slightly beyond the requirements of the syllabus in order to arrive at a satisfactory termination of that topic. The assessable learning outcomes addressed in each section, or group of sections, of a chapter are listed using OCR wording, so that students can identify the exact coverage of the syllabus.

Key points, definitions and equations are highlighted in yellow panels. After each section or group of sections there is a brief summary of the ground covered. Mathematical derivations have been kept to a minimum, but where some guidance may be helpful, this is provided in a *Maths Note* box. To achieve familiarity through practice, worked examples are provided at frequent intervals. These are followed by similar questions for students to attempt, under the heading *Now it's your turn*. In addition, there are groups of problems and questions at the end of each section or group of sections, and a selection of exam-style questions at the end of each chapter. Answers to these questions can be found at the back of the book.

The website has a series of Personal Tutor worked examples, which are audio-visual resources containing step by step instructions on how to complete mathematical aspects of the course, offering support to students when they are working on their own and allowing them to work at their own pace.

Chris Mee
Mike Crundell
Brian Arnold
Wendy Brown

Acknowledgements

The Publishers would like to thank the following for permission to reproduce copyright material:

Picture credits

p.6 Fig 1.2 Edward Kinsman/Science Photo Library; **p.6** Fig 1.3 Ping Europe Ltd; **p.22** Fig. 2.9 Tek Image/Science Photo Library; **p.22** Fig. 2.10 Novosti/ Science Photo Library; **p.23** Fig. 2.11 The Travel Library/Rex Features; **p.38** Fig. 2.27 Andrew Lambert Photography/Science Photo Library; **p.40** Fig. 2.32 Andrew Lambert Photography/Science Photo Library; **p.41** Fig. 2.33 Kit Kittle/Corbis; **p.42** Fig. 2.34 AP/PA Photos; **p.51** Fig. 3.2 David Ducros/Science Photo Library; **p.60** Fig. 4.5 Science Photo Library; **p.105** Fig. 6.1 Hodder Education; **p.107** Fig. 6.7 Science Photo Library; **p.109** Fig. 6.16 Pete Saloutos/Corbis; **p.117** Fig. 6.25 Lawrence Berkeley Laboratory/Science Photo Library; **p.120** Fig. 6.29 Steve Allen/Science Photo Library; **p.137** Fig. 6.46 GeoScience Features; **p.146** Fig. 7.2 N Feather/ Science Photo Library; **p.151** Fig. 7.5 Martyn F Chillmaid/Science Photo Library; **p.157** Fig. 7.11 Andrew Lambert Photography/Science Photo Library; **p.171** Fig. 7.17 European Space Agency/Science Photo Library; **p.172** Fig. 7.18 Prof Erwin Mueller/Science Photo Library; **p.175** Fig. 7.22 Dr David Wexler/Science Photo Library; **p.184** Fig. 8.1 Gustoimages/Science Photo Library; **p.192** Fig. 8.9 Sovereign, Ism/Science Photo Library; **p.193** Fig. 8.11 Sovereign, Ism/Science Photo Library; **p.194** Fig. 8.13 Mauro Fermariello/ Science Photo Library; **p.194** Fig. 8.14 Zephyr/Science Photo Library; **p.201** Fig. 8.21 Aj Photo/Hop Americain/Science Photo Library; **p.202** Fig. 8.23 Sovereign, Ism/Science Photo Library; **p.206** Fig. 8.27 RVI Medical Physics, Newcastle / Simon Fraser / Science Photo Library; **p.207** Fig. 8.28 BSIP, Raguet / Science Photo Library; **p.208** Fig. 8.29 Wellcome Dept. of Cognitive Neurology/Science Photo Library; **p.212** Fig. 8.31 Zephyr/Science Photo Library; **p.220** Fig. 9.4 Robert Harding Picture Library Ltd / Alamy; **p.224** Fig. 9.5 NASA/CXC/ESA/ASU/Science Photo Library

Newton's laws of motion and collisions

1.

Motion may be described in terms of displacement, velocity and acceleration. Now we will try to explain *why* bodies move. This will introduce Newton's three laws of motion, the fundamental laws which state and define what forces do in relation to motion. We shall then see how Newton's laws help us to explain what happens in collisions, which are important in many areas of physics. The total momentum of a system of colliding bodies remains constant. This leads us to the principle of conservation of momentum, one of the important conservation laws of physics.

1.1 Force and motion

At the end of Section 1.1 you should be able to:
- state and use each of Newton's three laws of motion
- explain that $F = ma$ is a special case of Newton's second law.

A full treatment of the concept of force is to be found in Chapter 4 of the AS book.

In general, a force is something that gives rise to either change of shape of an object or to change of velocity of an object. That is, it tends to cause the object to accelerate.

The relation between force and motion can be summarised in Newton's laws of motion.

Newton's first law

Every body stays in a state of rest or of constant velocity unless acted upon by a resultant force.

This law tells us that if we want to change the motion of a body, then a force is required.

Newton's second law

When a force acts on a body, then the force is proportional to the rate of change of momentum of the body in the direction of the force.

The force and the change of momentum are in the same direction.

Remember that momentum is equal to *mass × velocity*. So, for an object with constant mass, the rate of change of momentum will be equal to *mass × rate of change of velocity*. We know that rate of change of velocity is *acceleration* and so, Newton's second law is often written as

Force = mass × acceleration

and in symbols

F = ma

This simplified version of Newton's law (i.e. the mass is constant) provides the means by which force may be defined since mass and acceleration can be expressed in terms of the fundamental quantities.

One newton is the force required to give a mass of one kilogram an acceleration of one metre per second per second in the direction of the force.

Newton's third law

Sometimes written as

To every action, there is an equal and opposite reaction.

This statement should be used with caution. Remember that the third law deals with two bodies that are interacting. This brief statement does not make it clear on what the forces (the action and the reaction) are acting. A fuller and more precise statement of the third law is

Whenever one body exerts a force on another, then the second body exerts an equal but opposite force of the same kind on the first body.

Section 1.1 Summary

- The force of friction opposes motion.
- Newton's laws of motion define *force*. The laws are:
  First law: Every body continues in its state of rest, or of uniform motion in a straight line, unless acted upon by a force.
  Second law: Force is proportional to mass × acceleration or *F = ma*, where force *F* is in newtons, mass *m* is in kilograms and acceleration *a* is in metres per second per second.
  Third law: Whenever one body exerts a force on another, the second exerts an equal and opposite force of the same kind on the first.

Section 1.1 Questions

1 The net force on a car is 230 N and this produces an acceleration of 0.21 m s^{-2}. Calculate the mass of the car.

2 A spherical lump of play clay of mass 95 g is 2.5 m above ground level. It is dropped from rest and it hits the ground without rebounding. The play clay is deformed and is flattened by 0.54 cm.
Calculate:
(a) the speed of the play clay as it reaches the ground,

(b) the average acceleration of the play clay when it hits the ground,
(c) the average force exerted by the ground on the play clay.

3 Some forces act on an object. State and explain whether it is possible for the object to have zero acceleration.

4 Estimate the mass of an apple. What is the weight of this apple?

1.2 Momentum

At the end of Section 1.2 you should be able to:
- define linear momentum
- define net force on a body as equal to rate of change of momentum
- use the equation $F = \Delta p/\Delta t$
- define impulse of a force
- recall and use the equation *impulse = change in momentum*
- recall that the area under a graph of force against time is equal to impulse.

We shall now introduce a quantity called **momentum**, and see how Newton's laws may be related to it.

The momentum of a particle is defined as the product of its mass and its velocity.

In words

momentum = mass × velocity

and in symbols

$p = mv$

The unit of momentum is the unit of mass times the unit of velocity; that is, kg m s^{-1}. An alternative unit is the newton second (N s). Momentum, like velocity, is a vector quantity. Its complete name is **linear momentum**, to distinguish it from angular momentum, which does not concern us here.

Newton's first law states that every body continues in a state of rest, or of uniform motion in a straight line, unless acted upon by a force. We can express this law in terms of momentum. If a body maintains its uniform motion in a straight line, its momentum is unchanged. If a body remains at rest, again its momentum (zero) does not change. Thus, an alternative statement of the first law is that the momentum of a particle remains constant unless an external force acts on the particle. As an equation

$$p = \text{constant}$$

This is a special case, for a single particle, of a very important conservation law: the principle of conservation of momentum. This word 'conservation' here means that the quantity remains constant.

Newton's second law can also be expressed in terms of momentum. We already have it in a form which relates the force acting on a body to the product of the mass and the acceleration of the body. Remember that the acceleration of a body is the rate of change of its velocity. The product of mass and acceleration is just the mass times the rate of change of velocity. For a body of constant mass, this is just the same as the rate of change of (mass times velocity). But (mass times velocity) is momentum, so the product of mass and acceleration is identical with the rate of change of momentum. Thus, the second law may be stated as

The resultant force acting on a body is equal to the rate of change of its momentum.

This is the formal definition of the second law. Expressed in terms of symbols

$$F = ma = m(\Delta v/\Delta t) = \Delta(mv)/\Delta t = \Delta p/\Delta t$$

Continuing with the idea of force being equal to rate of change of momentum, the third law relating to action and reaction forces becomes: the rate of change of momentum due to the action force on one body is equal and opposite to the rate of change of momentum due to the reaction force on the other body. Simplifying this by considering a unit time interval, this becomes: when two bodies exert action and reaction forces on each other, their changes of momentum are equal and opposite.

1.3 Momentum and collisions

At the end of Section 1.3 you should be able to:
- state the principle of conservation of linear momentum
- apply the principle of conservation of momentum to solve problems in one dimension
- define what is meant by *elastic* and *inelastic* collisions
- explain that, whilst momentum of a system is always conserved in the interaction between bodies, some change in kinetic energy usually occurs.

Momentum and impulse

It is now useful to introduce a quantity called **impulse** and relate it to a change in momentum.

If a constant force F acts on a body for a time Δt, the impulse of the force is given by $F\Delta t$.

The unit of impulse is given by the unit of force, the newton, multiplied by the unit of time, the second: it is the newton second (N s).

We know from Newton's second law that the force acting on a body is equal to the rate of change of momentum of the body. We have already expressed this as the equation

$$F = \Delta p/\Delta t$$

If we multiply both sides of this equation by Δt, we obtain

$$F\Delta t = \Delta p$$

We have already defined $F\Delta t$ as the impulse of the force. The right-hand side of the equation (Δp) is the change in the momentum of the body. So, Newton's second law tells us that the **impulse of a force is equal to the change in momentum**. It is useful for dealing with forces that act over a short interval of time, as in a collision. The forces between colliding bodies are seldom constant throughout the collision, but the equation can be applied to obtain information about the average force acting.

Note that the idea of impulse explains why there is an alternative unit for momentum. In Section 1.2 we introduced the kg m s^{-1} and the N s as possible units for momentum. The kg m s^{-1} is the logical unit, the one you arrive at if you take momentum as being the product of mass and velocity. The N s comes from the impulse–momentum equation: it is the unit of impulse, and because impulse is equal to change of momentum, it is also a unit for momentum.

Example

Some tennis players can serve the ball at a speed of $55\,\text{m s}^{-1}$. The tennis ball has a mass of $60\,\text{g}$. In an experiment it is determined that the ball is in contact with the racket for $25\,\text{ms}$ during the serve (Figure 1.1). Calculate the average force exerted by the racket on the ball.

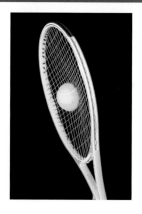

Figure 1.1

The change in momentum of the ball as a result of the serve is $0.060 \times 55 = 3.3\,\text{kg m s}^{-1}$. By the impulse–momentum equation, the change in momentum is equal to the impulse of the force. Since impulse is the product of force and time, $Ft = 3.3\,\text{N s}$.

Here t is $0.025\,\text{s}$; thus $F = 3.3/0.025 = \mathbf{132\,N}$.

Now it's your turn

Figure 1.2

A golfer hits a ball of mass $45\,\text{g}$ at a speed of $40\,\text{m s}^{-1}$ (Figure 1.2). The golf club is in contact with the ball for $3.0\,\text{ms}$. Calculate the average force exerted by the club on the ball.

The principle of conservation of momentum

Figure 1.3 System of two particles

We have already seen that Newton's first law states that the momentum of a single particle is constant, if no external force acts on the particle. Now think about a system of two particles (Figure 1.3). We allow these particles to exert some sort of force on each other: it could be gravitational attraction or, if the particles were charged, it could be electrostatic attraction or repulsion.

These two particles are isolated from the rest of the Universe, and experience no outside forces at all. If the first particle exerts a force F on the second, Newton's third law tells us that the second exerts a force $-F$ on the first. The minus sign indicates that the forces are in opposite directions. As we saw in the last section, we can express this law in terms of change of momentum. The change of momentum of the second particle as a result of the force exerted on it by the first is equal and opposite to the change of momentum of the first particle as a result of the force exerted on it by the second. Thus, the changes of momentum of the individual particles cancel out, and the momentum of the system of two particles remains constant. The particles have merely exchanged some momentum. The situation is expressed by the equation

$$p = p_1 + p_2 = \text{constant}$$

where p is the total momentum, and p_1 and p_2 are the individual momenta.

We could extend this idea to a system of three, four, or finally any number n of particles.

> If no external force acts on a system, the total momentum of the system remains constant, or is conserved.

A system on which no external force acts is often called an *isolated system*. The fact that the total momentum of an isolated system is constant is the **principle of conservation of momentum**. It is a direct consequence of Newton's third law of motion.

Collisions

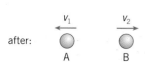

Figure 1.4 Collision between two particles

We now use the principle of conservation of momentum to analyse a system consisting of two colliding particles. (If you want a real example to think about, try billiard balls.)

Consider two particles A and B making a direct, head-on collision. Particle A has mass m_1 and is moving with velocity u_1 in the direction from left to right; B has mass m_2 and has velocity u_2 in the direction from right to left (Figure 1.4). As velocity is a vector quantity, this is the same as saying that the velocity is $-u_2$ from left to right. The particles collide. After the collision they have velocities $-v_1$ and v_2 respectively in the direction from left to right. That is, both particles are moving back along their directions of approach.

According to the principle of conservation of momentum, the total momentum of this isolated system remains constant, whatever happens as a result of the interaction of the particles. Thus, the total momentum before the collision must be equal to the total momentum after the collision. The momentum before the collision is

$$m_1 u_1 - m_2 u_2$$

and the momentum after is

$$-m_1 v_1 + m_2 v_2$$

Because total momentum is conserved

$$m_1 u_1 - m_2 u_2 = -m_1 v_1 + m_2 v_2$$

Knowing the masses of the particles and the velocities before collision, this equation would allow us to calculate the relation between the velocities after the collision.

The way to approach collision problems is as follows.

- Draw a labelled diagram showing the colliding bodies before collision. Draw a separate diagram showing the situation after the collision. Take care to define the directions of all the velocities.
- Obtain an expression for the total momentum before the collision, remembering that momentum is a vector quantity. Similarly, find the total momentum after the collision, taking the same reference direction.
- Then equate the momentum before the collision to the momentum afterwards.

Examples

1 What is the magnitude of the momentum of an α-particle of mass 6.6×10^{-27} kg travelling with a speed of 2.0×10^7 m s^{-1}?

$$p = mv = 6.6 \times 10^{-27} \times 2.0 \times 10^7 = \mathbf{1.3 \times 10^{-19} \, kg \, m \, s^{-1}}$$

2 A cannon of mass 1.5 tonnes (1.5×10^3 kg) fires a cannon-ball of mass 5.0 kg (Figure 1.5). The speed with which the ball leaves the cannon is 70 m s^{-1} relative to the Earth. What is the initial speed of recoil of the cannon?

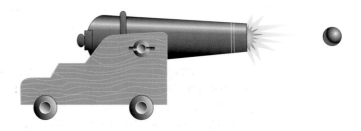

Figure 1.5

The system under consideration is the cannon and the cannon-ball. The total momentum of the system before firing is zero. Because the total momentum of an isolated system is constant, the total momentum after firing must also be zero. That is, the momentum of the cannon-ball, which is $5.0 \times 70 = 350$ kg m s^{-1} to the right, must be exactly balanced by the momentum of the cannon. If the initial speed of recoil is v, the momentum of the cannon is $1500v$ to the left. Thus, $1500v = 350$ and $v = \mathbf{0.23\,m\,s^{-1}}$.

Now it's your turn

1 What is the magnitude of the momentum of an electron of mass 9.1×10^{-31} kg travelling with a speed of 7.5×10^{6} m s^{-1}?

2 An ice-skater of mass 80 kg, initially at rest, pushes his partner, of mass 65 kg, away from him so that she moves with an initial speed of 1.5 m s^{-1}. What is the initial speed of the first skater after this manoeuvre?

The example of the tennis racket hitting the ball allowed us to calculate the average force between the racket and the ball. In fact, the force would increase from zero to a maximum value and then decrease back to zero. The variation with time of the force may be illustrated in Figure 1.6.

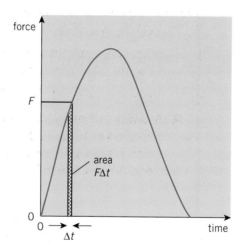

Figure 1.6

If it is assumed that the force F is constant for a time Δt, then the impulse provided by this force, $F\Delta t$, is represented by the area of the strip below the graph line on Figure 1.6.

When all these individual strips are added, then the area below the line of the *force–time* graph gives us the magnitude of the impulse. Since impulse is equal to change in momentum, then this area below the graph line is equal to the change in momentum.

Example

The graph of Figure 1.7 shows the variation with time of the force acting on a ball as it is hit by a racket.

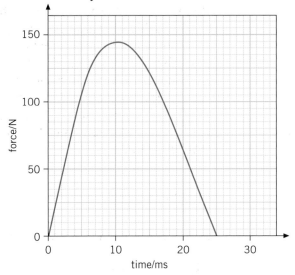

Figure 1.7

The ball has a mass of 60 g. Estimate the speed of the ball as it leaves the racket.

The area below the line is approximately 18 square units.
1.0 square unit represents $(25 \times 5.0 \times 10^{-3})$ N s = 0.125 N s
impulse acting on the ball = 18×0.125 = 2.25 N s
change in momentum of the ball = 2.25 kg m s^{-1}
change in momentum = mass $\times$ change in velocity
speed of ball = $2.25/(60 \times 10^{-3})$ = 38 m s^{-1}

Now it's your turn

The graph of Figure 1.8 shows the variation with time of the resultant force on a bullet as it moves down the barrel of a gun.

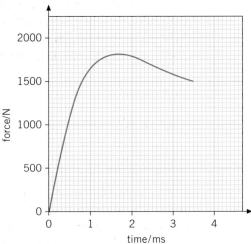

Figure 1.8

The bullet has a mass of 24 g.

Calculate the speed of the bullet as it leaves the barrel.

Elastic and inelastic collisions

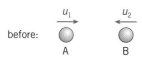

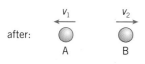

Figure 1.9 Collision between two particles

In some collisions, kinetic energy is conserved as well as momentum. By the **conservation of kinetic energy**, we mean that the total kinetic energy of the colliding bodies before collision is the same as the total kinetic energy afterwards. This means that no energy is lost in the permanent deformation of the colliding bodies, or as heat and sound. There is a transformation of energy during the collision: while the colliding bodies are in contact, some of the kinetic energy is transformed into elastic potential energy, but as the bodies separate, it is transformed into kinetic energy again.

Using the same notation for the masses and speeds of the colliding particles as in Section 1.3 (see Figure 1.9), the total kinetic energy of the particles before collision is

$$\tfrac{1}{2}m_1u_1{}^2 + \tfrac{1}{2}m_2u_2{}^2$$

The total kinetic energy afterwards is

$$\tfrac{1}{2}m_1v_1{}^2 + \tfrac{1}{2}m_2v_2{}^2$$

If the collision is **elastic**, the kinetic energy before collision is equal to the kinetic energy after collision.

$$\tfrac{1}{2}m_1u_1{}^2 + \tfrac{1}{2}m_2u_2{}^2 = \tfrac{1}{2}m_1v_1{}^2 + \tfrac{1}{2}m_2v_2{}^2$$

Note that because energy is a scalar quantity, the directions of motion of the particles are not indicated by the signs of the various terms.

In solving problems about elastic collisions, this equation is useful because it gives another relation between masses and velocities, in addition to that obtained from the principle of conservation of momentum.

When the velocity directions are as defined in Figure 1.9, application of the two conservation conditions shows that

$$u_1 + u_2 = v_1 + v_2$$

for a perfectly elastic collision. That is, the relative speed of approach $(u_1 + u_2)$ is equal to the relative speed of separation $(v_1 + v_2)$. Note that this useful relation applies *only for a perfectly elastic collision*.

Elastic collisions occur in the collisions of atoms and molecules. We shall see in Chapter 4 that one of the most important assumptions in the kinetic theory of gases is that the collisions of the gas molecules with the walls of the container are perfectly elastic. However, in larger scale collisions, such as those of billiard balls, collisions cannot be perfectly elastic. (The 'click' of billiard balls on impact indicates that a very small fraction of the total energy of the system has been transformed into sound.) Nevertheless, we often make the assumption that such a collision is perfectly elastic.

Collisions in which the total kinetic energy is not the same before and after the event are called **inelastic**.

Total energy must, of course, be conserved. But in an inelastic collision the kinetic energy that does not re-appear in the same form is transformed into heat, sound and other forms of energy. In an extreme case, all the kinetic energy may be lost. A lump of modelling clay dropped on to the floor does not bounce. All the kinetic energy it possessed just before hitting the floor has been transformed into the work done in flattening the lump, and (a much smaller amount) into the sound energy emitted as a 'squelch'.

Although kinetic energy may or may not be conserved in a collision, momentum is always conserved, and so is total energy.

The truth of this statement may not be entirely obvious, especially when considering examples such as the lump of modelling clay which was dropped on to the floor. Surely the clay had momentum just before the collision with the floor, but had no momentum afterwards? True! But for the system of the lump of clay alone, external forces (the attraction of the Earth on the clay, and the force exerted by the floor on the clay on impact) were acting. When external forces act, the conservation principle does not apply. We need to consider a system in which no external forces act. Such a system is the lump of modelling clay and the Earth. While the clay falls towards the floor, gravitational attraction will also pull the Earth towards the clay. Conservation of momentum can be applied in that the total momentum of clay and Earth remains constant throughout the process: before the collision, and after it. The effects of the transfer of the clay's momentum to the Earth are not noticeable due to the difference in mass of the two objects.

Example

A billiard ball A moves with speed u_A directly towards a similar ball B which is at rest (Figure 1.10). The collision is perfectly elastic. What are the speeds v_A and v_B after the collision?

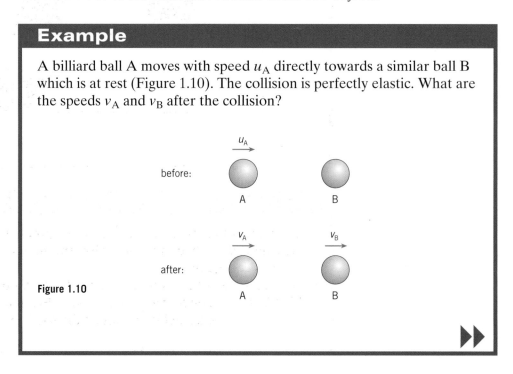

Figure 1.10

It is convenient to take the direction from left to right as the direction of positive momentum. If the mass of a billiard ball is m, the total momentum of the system before the collision is mu_A. By the principle of conservation of momentum, the total momentum after collision is the same as that before, or

$$mu_A = mv_A + mv_B$$

The collision is perfectly elastic, so the total kinetic energy before the collision is the same as that afterwards, or

$$\tfrac{1}{2}mu_A{}^2 = \tfrac{1}{2}mv_A{}^2 + \tfrac{1}{2}mv_B{}^2$$

Solving these equations gives $v_A = 0$ and $v_B = u_A$. That is, ball A comes to a complete standstill, and ball B moves off with the same speed as that with which ball A struck it. (Another solution is possible algebraically: $v_A = u_A$ and $v_B = 0$. This corresponds to a non-collision. Ball A is still moving with its initial speed, and ball B is still at rest. In cases where algebra gives us two possible solutions, we need to decide which is the one that is physically appropriate.)

Now it's your turn

A trolley A moves with speed u_A towards a trolley B of equal mass which is at rest (Figure 1.11). The trolleys stick together and move off as one with speed $v_{A,B}$.
(a) Find $v_{A,B}$.
(b) What fraction of the initial kinetic energy of trolley A is converted into other forms in this inelastic collision?

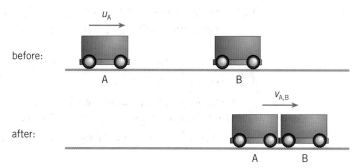

Figure 1.11

Sections 1.2–1.3 Summary

■ The linear momentum p of a body is the product of its mass m and its velocity v, described as $p = mv$. Momentum has units kg m s^{-1} or N s. It is a vector quantity.
■ Newton's laws of motion can be stated in terms of momentum.
   First law: The momentum of a particle remains constant unless an external force acts on the particle: p = constant

Second law: The resultant force acting on a body is equal to the rate of change of its momentum: $F = \Delta p/\Delta t$

Third law: When two bodies exert action and reaction forces on each other, their changes of momentum are equal and opposite.

■ The area below the line of a force–time graph represents the impulse or the change in momentum.

■ The principle of conservation of momentum states that the total momentum of an isolated system is constant. An isolated system is one on which no external force acts.

■ In collisions between bodies, application of the principle of conservation of momentum shows that the total momentum of the system before the collision is equal to the total momentum after the collision.

■ An elastic collision is one in which the total kinetic energy remains the same. In this situation, the relative speed of approach is equal to the relative speed of separation.

■ An inelastic collision is one in which the total kinetic energy is not the same before and after the event.

■ Although kinetic energy may or may not be conserved in a collision, momentum is always conserved, and so is total energy.

■ The impulse of a force F is the product of the force and the time Δt for which it acts:

$$\text{impulse} = F\Delta t$$

■ The impulse of a force acting on a body is equal to the change of momentum of the body:

$$F\Delta t = \Delta p$$

■ The unit of impulse is N s.

Sections 1.2–1.3 Questions

1 Calculate the magnitude of the momentum of a car of mass 1.5 tonnes (1.5×10^3 kg) travelling at a speed of 22 m s^{-1}.

2 When a certain space rocket is taking off, the propellant gases are expelled at a rate of 900 kg s^{-1} and speed of 40 km s^{-1}. Calculate the thrust on the rocket.

3 An insect of mass 4.5 mg, flying with a speed of 0.12 m s^{-1}, encounters a spider's web, which brings it to rest in 2.0 ms. Calculate the average force exerted by the insect on the web.

4 An atomic nucleus at rest emits an α-particle of mass 4 u. The speed of the α-particle is found to be 5.6×10^6 m s^{-1}. Calculate the speed with which the daughter nucleus, of mass 218 u, recoils.

5 A heavy particle of mass m_1, moving with speed u, makes a head-on collision with a light particle of mass m_2, which is initially at rest. The collision is perfectly elastic, and m_2 is very much less than m_1. Describe the motion of the particles after the collision.

6 A light body and a heavy body have the same momentum. Which has the greater kinetic energy?

Exam-style Questions

1 A bullet of mass $12\,\text{g}$ is fired horizontally from a gun with a velocity of $180\,\text{m s}^{-1}$. It hits, and becomes embedded in, a block of wood of mass $2000\,\text{g}$, which is freely suspended by long strings, as shown in Figure 1.12.

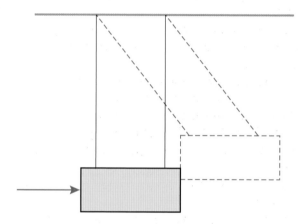

Figure 1.12

Calculate:
(a) (i) the magnitude of the momentum of the bullet just before it enters the block,
(ii) the magnitude of the initial velocity of the block and bullet after impact,
(iii) the kinetic energy of the block and embedded bullet immediately after the impact.
(b) Deduce the maximum height above the equilibrium position to which the block and embedded bullet rise after impact.

2 A nucleus A of mass $222\,\text{u}$ is moving at a speed of $350\,\text{m s}^{-1}$. While moving, it emits an α-particle of mass $4\,\text{u}$. After the emission, it is determined that the daughter nucleus, of mass $218\,\text{u}$, is moving with speed $300\,\text{m s}^{-1}$ in the original direction of the parent nucleus. Calculate the speed of the α-particle.

3 A safety feature of modern cars is the air-bag, which, in the event of a collision, inflates and is intended to decrease the risk of serious injury. Use the concept of impulse to explain why an air-bag might have this effect.

4 Two frictionless trolleys A and B, of mass m and $3m$ respectively, are on a horizontal track (Figure 1.13). Initially they are clipped together by a device which incorporates a spring, compressed between the trolleys. At time $t = 0$ the clip is released. The velocity of trolley B is u to the right.

Figure 1.13

(a) Calculate the velocity of trolley A as the trolleys move apart.
(b) At time $t = t_1$, trolley A collides elastically with a fixed spring and rebounds. (Compression and expansion of the spring take a negligibly short time.) Trolley A catches up with trolley B at time $t = t_2$.
   (i) Calculate the velocity of trolley A between $t = t_1$ and $t = t_2$.
   (ii) Find an expression for t_2 in terms of t_1.
(c) When trolley A catches up with trolley B at time t_2 the clip operates so as to link them again, this time without the spring between them, so that they move together with velocity v. Calculate the common velocity v in terms of u.
(d) Initially the trolleys were at rest and the total momentum of the system was zero. However, the answer to (c) shows that the total momentum after $t = t_2$ is not zero. Discuss this result with reference to the principle of conservation of momentum.

5 A ball of mass m makes a perfectly elastic head-on collision with a second ball, of mass M, initially at rest. The second ball moves off with half the original speed of the first ball.
(a) Express M in terms of m.
(b) Determine the fraction of the original kinetic energy retained by the ball of mass m after the collision.

2. Circular motion and oscillations

In the AS book we dealt with uniformly accelerated motion. In this chapter we shall study two important examples of motion in which the acceleration is not uniform – circular motion and simple harmonic oscillations. Both have very wide applications in physics and in everyday life. Understanding the physics of circular motion, for example, has enabled scientists to launch satellites into orbit around the Earth and to design theme park rides.

2.1 Going round in circles

At the end of Section 2.1 you should be able to:
- define the radian
- convert angles from degrees to radians and *vice versa*
- explain that a force perpendicular to the velocity of an object will make the object describe a circular path
- explain what is meant by centripetal acceleration and centripetal force
- select and apply equations for speed $v = 2\pi r/T$ and centripetal acceleration $a = v^2/r$
- select and apply the equation for centripetal force $F = ma = mv^2/r$.

Radians and degrees

In circular motion, it is convenient to measure angles in **radians** rather than degrees. One degree is by tradition equal to the angle of a circle divided by 360.

> One radian (rad) is defined as the angle subtended at the centre of a circle by an arc equal in length to the radius.

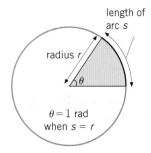

Figure 2.1 θ in radians = arc/radius

Thus, to obtain an angle in radians, we divide the length of the arc by the radius of the circle (see Figure 2.1).

$$\theta = \frac{\text{length of arc}}{\text{radius of circle}}$$

The angle in radians in a complete circle would be

$$\theta = \frac{\text{circumference of the circle}}{\text{radius}} = \frac{2\pi r}{r} = 2\pi$$

Since the angle of a complete circle is 360°, then

$$2\pi \text{ rad} = 360°$$

or

$$1 \text{ rad} = 57.3°$$

Angular velocity

For an object moving in a circle:

> The **angular speed** is defined as the angle swept out by the radius per unit time.

The **angular velocity** is the angular speed in a given direction (for example, clockwise). The unit of angular speed and angular velocity is the radian per second (rad s^{-1}).

$$angular \; speed \; \omega = \frac{\Delta\theta}{\Delta t}$$

Figure 2.2 shows an object travelling at constant speed v in a circle of radius r. In a time Δt the object moves along an arc of length Δs and sweeps out an angle $\Delta\theta$. From the definition of the radian,

$$\Delta\theta = \Delta s/r \quad \text{or} \quad \Delta s = r\Delta\theta$$

Dividing both sides of this equation by Δt,

$$\Delta s/\Delta t = r\Delta\theta/\Delta t$$

By definition, $\Delta s/\Delta t$ is the linear speed v of the object, and $\Delta\theta/\Delta t$ is the angular speed ω. Hence,

$$v = r\omega$$

Remember that if an object com\pletes one revolution of a circle radius r in a time T, then the linear speed v is given by $v = \dfrac{\text{circumference}}{\text{time}}$

$$v = \frac{2\pi r}{T}$$

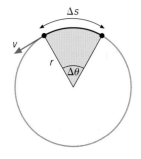

Figure 2.2 Angular speed $\omega = v/r$

Centripetal force

Newton's first law of motion (see Chapter 1) tells us that an object with a resultant force of zero acting on it will either not be moving at all, or will be moving in a straight line at constant speed (that is, its velocity does not change). The object is said to be in equilibrium. (The full conditions of equilibrium require there to be no resultant force or resultant moment acting on the body.)

An object travelling in a circle may have a constant speed, but it is not travelling in a straight line. Therefore its velocity is changing, since velocity is speed in a certain direction. A change in velocity means the object is accelerating.

In the AS book we dealt with objects moving along a straight line, where acceleration related to changes of speed. In circular motion, objects accelerate because their *direction* changes. The acceleration a of an object moving in a circle of radius r with speed v is given by the expression

$$a = \frac{v^2}{r}$$

We shall derive this later. This acceleration is towards the centre of the circle. It is called the **centripetal acceleration**. In order to make an object accelerate, there must be a resultant force acting on it. This force is called the **centripetal force**. From Newton's second law of motion (see Chapter 1), the force F to give to mass m an acceleration a is

$$F = ma$$

By substituting for a, the centripetal force is

$$F = \frac{mv^2}{r}$$

The centripetal force acts towards the centre of the circle.

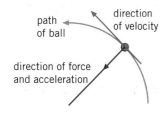

Figure 2.3 A ball swung in a circle on the end of a string

This means that the centripetal force acts at right angles to the instantaneous velocity of the object.

Consider a ball on a string which is being swung in a horizontal circle. The tension in the string provides the centripetal force. At any instant the direction of the ball's velocity is along the tangent to the circle, as shown in Figure 2.3. If the string breaks or is released, there is no longer any centripetal force. The ball will travel in the direction of the tangent to the circle at the moment of release.

Example

An aircraft in a display team makes a turn in a horizontal circle of radius 500 m. It is travelling at a speed of 100 m s^{-1}.
(a) Calculate the centripetal acceleration of the aircraft.
(b) The mass of the pilot in the aircraft is 80 kg. Calculate the centripetal force acting on the pilot.

(a) From $a = v^2/r$, the centripetal acceleration $a = 100^2/500 =$ **20 m s^{-2}**.
(b) From $F = ma$, the centripetal force on the pilot $F = 80 \times 20 =$ **1600 N**.

Now it's your turn

1 In a model of the hydrogen atom, the electron moves round the nucleus in a circular orbit of radius 5.3×10^{-11} m. The speed of the electron is 2.2×10^6 m s^{-1}. Calculate the centripetal acceleration of the electron.

2 A satellite orbits the Earth 200 km above its surface. The acceleration towards the centre of the Earth is 9.2 m s^{-2}. The radius of the Earth is 6400 km. Calculate:
   (a) the speed of the satellite,
   (b) the time to complete one orbit.

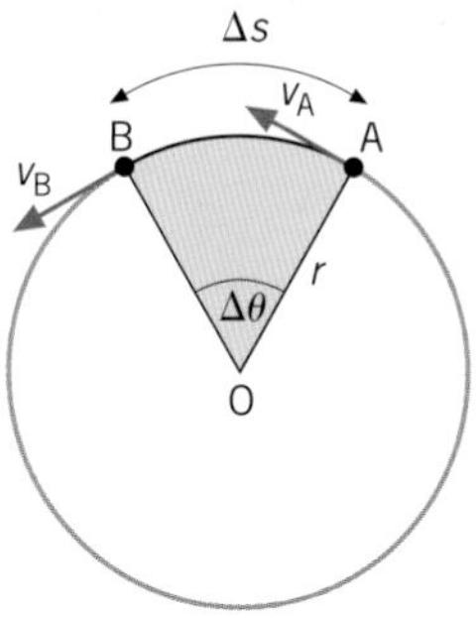

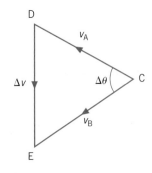

Figure 2.4 Diagram for proof of $a = v^2/r$

Derivation of expression for centripetal acceleration

Figure 2.4 shows an object which has travelled at constant speed v in a circular path from A to B in time Δt. At A, its velocity is v_A, and at B the velocity is v_B. Both v_A and v_B are vectors.

The change in velocity Δv may be seen in the vector diagram of Figure 2.5. A vector Δv must be added to v_A in order to give the new velocity v_B.

The angle between the two radii OA and OB is $\Delta\theta$. This angle is also equal to the angle between the vectors v_A and v_B, because triangles OAB and CDE are similar. Consider angle $\Delta\theta$ to be so small that the arc AB may be approximated to a straight line. Then, using similar triangles, DE/CD = AB/OA, and $\Delta v/v_A = \Delta s/r$ or

$$\Delta v = \Delta s(v_A/r)$$

The time to travel either the distance Δs or the angle $\Delta\theta$ is Δt. Dividing both sides of the equation by Δt,

$$\Delta v/\Delta t = (\Delta s/\Delta t)(v/r)$$

and from the definitions of acceleration ($a = \Delta v/\Delta t$) and speed ($v = \Delta s/\Delta t = v_A = v_B$) we have

$$a = v(v/r) \quad \text{or} \quad a = v^2/r$$

Figure 2.5 Vector diagram for proof of $a = v^2/r$

This expression can be written in terms of angular speed ω. Since $v = r\omega$,

$$\text{centripetal acceleration} = \frac{v^2}{r} = r\omega^2$$

and, since $F = ma$,

$$\text{centripetal force} = \frac{mv^2}{r} = mr\omega^2$$

Example

The drum of a spin dryer has a radius of 20 cm and rotates at 600 revolutions per minute.
(a) Show that the angular speed of the drum is about 63 rad s^{-1}.
(b) Calculate, for a point on the edge of the drum:
   (i) its linear speed,
   (ii) its acceleration towards the centre of the drum.

(a) 600 revolutions per minute is 10 revolutions per second. The time for one revolution is thus 0.10 s. Each revolution is 2π rad, so the angular speed $\omega = \theta/t = 2\pi/0.10 =$ **63 rad s^{-1}**.
(b) (i) Using $v = r\omega$, $v = 0.20 \times 63 =$ **13 m s^{-1} (12.6 m s^{-1})**.
     OR

$$\text{using } v = \frac{2\pi r}{T}$$

$$\text{time } T \text{ for one revolution} = \frac{60}{600} \text{ s}$$

$$\text{so } v = \frac{2\pi \times 20}{\frac{1}{10}}$$

$$= \textbf{1260 cm s}^{-1}$$
$$\simeq \textbf{13 m s}^{-1}$$

   (ii) Using $a = v^2/r$, $a = (13)^2/0.20 =$ **800 m s^{-2}**.

Now it's your turn

A toy train moves round a circular track of diameter 0.70 m, completing one revolution in 10 seconds. Calculate, for this train:
(a) the linear speed,
(b) the angular speed,
(c) the centripetal acceleration.

Examples of circular motion

When a ball is whirled round on the end of a string, you can see clearly that the string is making the ball accelerate towards the centre of the circle. However, in other examples it is not always so easy to see what force is providing the centripetal acceleration.

A satellite in Earth orbit experiences gravitational attraction towards the centre of the Earth. This attractive force provides the centripetal force and causes the satellite to accelerate towards the centre of the Earth, and so it moves in a circle. We shall return to this in detail in Chapter 3.

A charged particle moving at right-angles to a magnetic field experiences a force at right-angles to its direction of motion, and therefore moves in the arc of a circle. This will be considered in more detail in Chapter 6.

For a car travelling in a curved path, the frictional force between the tyres and the road surface provides the centripetal force. If this frictional force is not large enough, for example if the road is oily and slippery, then the car carries on moving in a straight line – it skids.

A passenger in a car that is cornering appears to be flung away from the centre of the circle. The centripetal force required to maintain the passenger in circular motion is provided through the seat of the car. This force is below the centre of mass of the passenger, causing rotation about the centre of mass (Figure 2.6). The effect is that the upper part of the passenger moves outwards unless another force acts on the upper part of the body, preventing rotation.

rotation about M

centre of mass M

force towards centre of circle

Figure 2.6 Passenger in a car rounding a corner

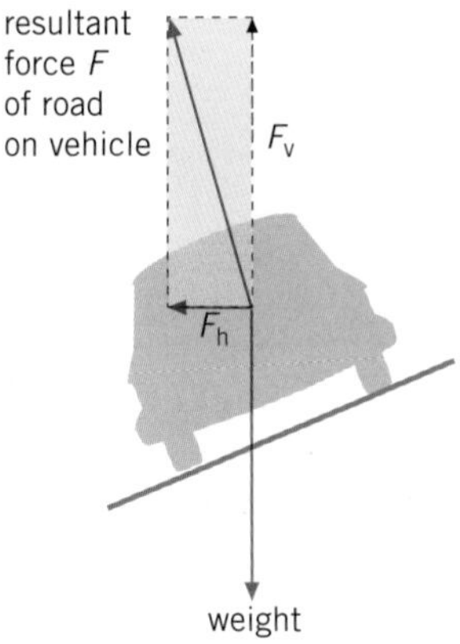

resultant force F of road on vehicle

F_v

F_h

weight

Figure 2.7 Cornering on a banked track

For cornering which does not rely only on friction, the road can be banked (see Figure 2.7). The road provides a resultant force normal to its surface through contact between the tyres and the road. This resultant force F is at an angle to the vertical, and can be resolved into a vertical component F_v and a horizontal component F_h. F_v is equal to the weight of the vehicle, thus maintaining equilibrium in the vertical direction. The horizontal component F_h provides the centripetal force towards the centre of the circle. Many roads are banked for greater road safety, so as to reduce the chance of loss of control of vehicles due to skidding outwards on the corner, and for greater passenger comfort.

An aircraft has a lift force caused by the different rates of flow of air above and below the wings. The lift force balances the weight of the aircraft when it flies on a straight, level path (Figure 2.8a). In order to change direction, the aircraft is banked so that the wings are at an angle to the horizontal (Figure 2.8b). The lift force now has a horizontal component which provides a centripetal force to change the aircraft's direction.

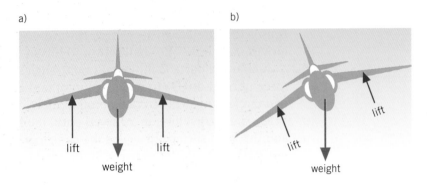

Figure 2.8 An aircraft a) in straight, level flight and b) banking

A centrifuge (Figure 2.9) is a device used to spin objects at high speed about an axis. It is used to separate particles in mixtures. More massive particles require larger centripetal forces in order to maintain circular motion than do less massive ones. As a result, the more massive particles tend to separate from less massive particles, collecting further away from the axis of rotation. Space research centres, such as NASA, use centrifuges which are large enough to rotate a person (Figure 2.10). Their purpose is to investigate the effects of large accelerations on the human body.

Figure 2.9 Separation of a solid from a liquid in a laboratory centrifuge

Figure 2.10 Centrifuge testing the effect of acceleration on the human body

Motion in a vertical circle

Some theme park rides involve rotation in a vertical circle. A person on such a ride must have a resultant force acting towards the centre of the circle.

The forces acting on the person are the person's weight, which always acts vertically downwards, and the reaction from the seat, which acts at right angles to the seat.

Figure 2.11 Ferris wheel at a theme park

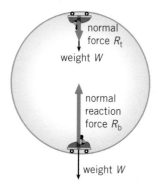

Figure 2.12 Forces on a person on a circular ride

Consider a person moving round a vertical circle at speed v. At the bottom of the ride, the reaction R_b from the seat must provide the centripetal force as well as overcoming the weight W of the person. Figure 2.12 illustrates the situation. Hence the centripetal force is given by

$$mv^2/r = R_b - W$$

At the top of the ride, the weight W and the reaction force R_t both act downwards towards the centre of the circle. The centripetal force is now given by

$$mv^2/r = R_t + W$$

This means that the force R_t from the seat at the top of the ride is less than the force R_b at the bottom. If the speed v is not large, then at the top of the circle the weight may be greater than the centripetal force. The person would lose contact with the seat and fall inwards.

Example

A rope is tied to a bucket of water, and the bucket is swung in a vertical circle of radius 1.2 m. What must be the minimum speed of the bucket at the highest point of the circle if the water is to stay in the bucket throughout the motion?

▶▶

This example is similar to the problem of the theme park ride. Water will fall out of the bucket if its weight is greater than the centripetal force. The critical speed v is given by $mv^2/r = mg$ or $v^2 = gr$. Here $v = \sqrt{(9.8 \times 1.2)} = \mathbf{3.4\,m\,s^{-1}}$.

Now it's your turn

At an air show, an aircraft diving at a speed of $170\,m\,s^{-1}$ pulls out of the dive by moving in the arc of a circle at the bottom of the dive.
(a) Calculate the minimum radius of this circle if the centripetal acceleration of the aircraft is not to exceed five times the acceleration of free fall.
(b) The pilot has mass $85\,kg$. What is the resultant force upwards on him at the instant when the aircraft is at its lowest point?

Section 2.1 Summary

- A resultant force acting towards the centre of the circle, called the centripetal force, is required to make an object move in a circle.
- Angles may be measured in radians (rad). One radian is the angle subtended at the centre of a circle by an arc of the circle equal in length to its radius.
- Angular speed ω is the angle swept out per unit time by a line rotating about a point.
- A particle moving along a circle of radius r with linear speed v has angular velocity ω given by: $v = r\omega$
- An object moving along a circle of radius r with linear speed v and angular speed ω has an acceleration a towards the centre (the centripetal acceleration) given by: $a = v^2/r = r\omega^2$
- For an object of mass m moving along a circle of radius r with linear speed v and angular speed ω, the centripetal force F is given by: $F = mv^2/r = mr\omega^2$

Section 2.1 Questions

1 State how the centripetal force is provided in the following examples:
(a) a planet orbiting the Sun,
(b) a child on a playground roundabout,
(c) a train on a curved track,
(d) a passenger in a car going round a corner.

2 NASA's 20-G centrifuge is used for testing space equipment and the effect of acceleration on humans. The centrifuge consists of an arm of length $17.8\,m$, rotating at constant speed and producing an acceleration equal to 20 times the acceleration of free fall. Calculate:
(a) the angular speed required to produce a centripetal acceleration of $20g$,
(b) the rate of rotation of the arm.
$(g = 9.8\,m\,s^{-2})$

2.2 Oscillations

At the end of Section 2.2 you should be able to:
- describe simple examples of free oscillations
- define and use the terms *displacement, amplitude, period, frequency* and *angular frequency*
- select and use the equation *period = 1/frequency*

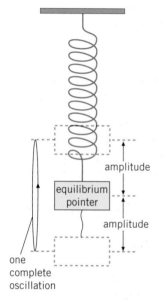

Figure 2.13 Oscillation of a mass on a spring

Some movements involve repetitive to-and-fro motion, such as a pendulum, the beating of a heart, the motion of a child on a swing, and the vibrations of a guitar string. Another example would be a mass bouncing up and down on a spring, as illustrated in Figure 2.13. One complete movement from the starting or rest position, up, then down and finally back up to the rest position, is known as an **oscillation**.

The time taken for one complete oscillation is referred to as the **period T** of the oscillation.

The oscillations repeat themselves.

The number of oscillations per unit time is the **frequency** f.

Frequency may be measured in hertz (Hz), where one hertz is one oscillation per second ($1\,\text{Hz} = 1\,\text{s}^{-1}$). However, frequency may also be measured in min^{-1}, hour^{-1}, etc. For example, it would be appropriate to measure the frequency of the tides in h^{-1}.

Since period T is the time for one oscillation then

frequency $f = 1/T$

As the mass oscillates, it moves from its rest position.

The distance from the equilibrium position is known as the **displacement**.

This is a vector quantity, since the displacement may be on either side of the equilibrium position.

The **amplitude** (a scalar quantity) is the maximum displacement.

Some oscillations maintain a constant period even when the amplitude of the oscillation changes. Galileo discovered this fact for a pendulum. He timed the swings of an oil lamp in Pisa Cathedral, using his pulse as a measure of time. Oscillators that have a constant time period are called isochronous, and may be made use of in timing devices. For example, in quartz watches the oscillations of a small quartz crystal provide constant time intervals. Galileo's experiment was not precise, and we now know that a pendulum swinging with a large amplitude is not isochronous.

The quantities period, frequency, displacement and amplitude should be familiar from our study of waves in the AS book. It should not be a surprise to meet them again, as the idea of oscillations is vital to the understanding of waves.

Displacement–time graphs

It is possible to plot displacement–time graphs for oscillators. One experimental method is illustrated in Figure 2.14. A mass on a spring oscillates above a position sensor that is connected to a computer through a datalogging interface causing a trace to appear on the monitor.

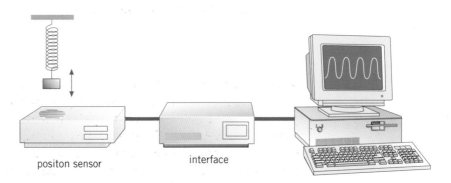

positon sensor   interface

Figure 2.14 Apparatus for plotting displacement–time graphs for a mass on a spring

The graph describing the variation of displacement with time may have different shapes, depending on the oscillating system. For many oscillators the graph is approximately a sine (or cosine) curve. A sinusoidal displacement–time graph is a characteristic of an important type of oscillation called **simple harmonic motion** (s.h.m.). Oscillators which move in s.h.m. are called **harmonic** oscillators. We shall analyse simple harmonic motion in some detail, because it successfully describes many oscillating systems, both in real life and in theory. Fortunately, the mathematics of s.h.m. can be approached through a simple defining equation. The properties of the motion can be deduced from the relations between graphs of displacement against time, and velocity against time, which we have already met in Chapter 2 of the AS book.

2.3 Simple harmonic motion

At the end of Section 2.3 you should be able to:

- define *simple harmonic motion*
- select and apply the equation $a = -(2\pi f)^2 x$ as the defining equation of simple harmonic motion
- select and use $x = A\cos(2\pi ft)$ or $x = A\sin(2\pi ft)$ as solutions to the equation $a = -(2\pi f)^2 x$
- select and apply the equation $v_{max} = (2\pi f)A$ for the maximum speed of a simple harmonic oscillator
- explain that the period of a simple harmonic oscillator is independent of its amplitude
- describe, with graphical illustrations, the changes in displacement, velocity and acceleration during simple harmonic motion.

Simple harmonic motion is defined as the motion of a particle about a fixed point such that its acceleration a is proportional to its displacement x from the fixed point, and is directed towards the point.

Mathematically, we write this definition as

$$a = -\omega^2 x$$

where ω^2 is a constant. We take the constant as a squared quantity, because this will ensure that the constant is always positive (the square of a positive number, or of a negative number, will always be positive). Why worry about keeping the constant positive? This is because the minus sign in the equation must be preserved. It has a special significance, because it tells us that the acceleration a is always in the opposite direction to the displacement x. Remember that both acceleration and displacement are vector quantities, so the minus sign is shorthand for the idea that the acceleration is always directed towards the fixed point from which the displacement is measured. This is illustrated in Figure 2.15.

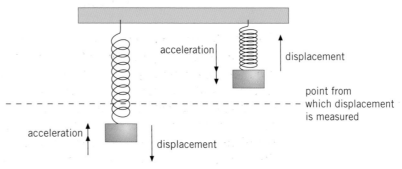

Figure 2.15 Directions of displacement and acceleration are always opposite

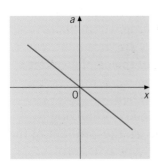

Figure 2.16 Graph of the defining equation for simple harmonic motion

The defining equation is represented in a graph of a against x as a straight line, of negative gradient, through the origin, as shown in Figure 2.16.

The gradient is negative because of the minus sign in the equation. Note that both positive and negative values for the displacement should be considered.

The square root of the constant ω^2 (that is, ω) is known as the **angular frequency** of the oscillation. This angular frequency ω is related to the frequency f of the oscillation by the expression

$$\omega = 2\pi f$$

The defining equation for s.h.m. may therefore be written as

$$a = -(2\pi f)^2 x$$

By Newton's second law, the force acting on a body is proportional to the acceleration of the body. The defining equation for simple harmonic motion can thus be related to the force acting on the particle. If the acceleration of the particle is proportional to its displacement from a fixed point, the resultant force acting on the particle is also proportional to the displacement. We can bring in the idea of the direction of the acceleration by specifying that the force is always acting towards the fixed point, or by calling it a **restoring** force.

Solution of equation for simple harmonic motion

In order to find the displacement–time relation for a particle moving in a simple harmonic motion, we need to solve the equation $a = -(2\pi f)^2 x$. To derive the solution requires mathematics which is beyond the requirements of A/AS Physics. However, you need to know the form of the solution. This is

$$x = A \sin 2\pi ft$$

or

$$x = A \cos 2\pi ft$$

where A is the amplitude of the oscillation. The solution $x = A \sin 2\pi ft$ is used when, at time $t = 0$, the particle is at its equilibrium position where $x = 0$. Conversely, if at time $t = 0$ the particle is at its maximum displacement, $x = A$, the solution is $x = A \cos 2\pi ft$. The variation with time t of the displacement x for the two solutions is shown in Figure 2.17.

In Chapter 2 of the AS book it was shown that the gradient of a displacement–time graph may be used to determine velocity. Referring to Figure 2.17, it can be seen that at each time at which $x = A$, the gradient of the graph is zero (this applies to both solutions). Thus, the velocity is zero whenever the particle has its maximum displacement. If we think about a mass vibrating up and down on a spring, this means that when the spring is fully stretched and the mass has its maximum displacement, the mass stops

a)
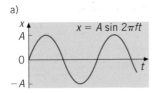

b)
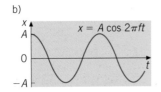

Figure 2.17 Displacement–time curves for the two solutions to the s.h.m. equation

a)

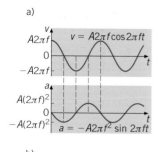

b)

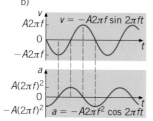

Figure 2.18 Velocity–time and acceleration–time graphs for the two solutions to the s.h.m. equation

moving downwards and has zero velocity. Also from Figure 2.17, we can see that the gradient of the graph is at a maximum whenever $x = 0$. This means that when the spring is neither under or over-stretched the speed of the mass is at a maximum. After passing this point the spring forces the mass to slow down until it changes direction.

If a full analysis is carried out, it is found that the variation of velocity with time is cosinusoidal if the sinusoidal displacement solution is taken, and sinusoidal if the cosinusoidal displacement solution is taken. This is illustrated in Figure 2.18.

The velocity v of the particle is given by the expressions

$$v = A2\pi f \cos 2\pi ft \quad \text{when} \quad x = A \sin 2\pi ft$$

and

$$v = -A2\pi f \sin 2\pi ft \quad \text{when} \quad x = A \cos 2\pi ft$$

In each case, there is a phase difference between velocity and displacement. The velocity curve is $\pi/2$ rad ahead of the displacement curve. (If the phase angle is not considered, the variation with time of the velocity is the same in each case.) The maximum speed v_{max} is given by

$$v_{\text{max}} = A2\pi f$$

There is an alternative expression for the velocity.

$$v = \pm 2\pi f \sqrt{(A^2 - x^2)}$$

which is derived on pages 34–35.

For completeness, Figure 2.18 also shows the variation with time of the acceleration a of the particle. This could be derived from the velocity–time graph by taking the gradient. The equations for the acceleration are

$$a = -A(2\pi f)^2 \sin 2\pi ft \quad \text{when} \quad x = A \sin 2\pi ft$$

and

$$a = -A(2\pi f)^2 \cos 2\pi ft \quad \text{when} \quad x = A \cos 2\pi ft$$

Note that, for both solutions, these equations are consistent with the defining equation for simple harmonic motion, $a = -(2\pi f)^2 x$. You can easily prove this by eliminating $\sin 2\pi ft$ from the first set of equations, and $\cos 2\pi ft$ from the second.

Example

The displacement x at time t of a particle moving in simple harmonic motion is given by $x = 0.25 \cos 7.5t$, where x is in metres and t is in seconds.

(a) Use the equation to find the amplitude, frequency and period for the motion.

(b) Find the displacement when $t = 0.50\,\mathrm{s}$.

(a) Compare the equation with $x = A \cos 2\pi ft$. The amplitude $A = \mathbf{0.25\,m}$. Remember that $\omega = 2\pi f$, so the frequency $f = 7.5/2\pi = \mathbf{1.2\,Hz}$. The period $T = 1/f = 1/1.2 = \mathbf{0.84\,s}$.

(b) Substitute $t = 0.50\,\mathrm{s}$ in the equation, remembering that the angle $2\pi ft$ is in radians and not degrees. $2\pi ft = 7.5 \times 0.50 = 3.75\,\mathrm{rad} = 215°$. So $x = 0.25 \cos 215° = \mathbf{-0.20\,m}$.

Now it's your turn

A mass oscillating on a spring has an amplitude of $0.10\,\mathrm{m}$ and a period of $2.0\,\mathrm{s}$.

(a) Deduce the equation for the displacement x if timing starts at the instant when the mass has its maximum displacement.

(b) Calculate the time interval from $t = 0$ before the displacement is $0.08\,\mathrm{m}$.

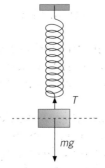

Figure 2.19 Mass on a helical spring

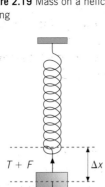

Figure 2.20 Mass on a helical spring: additional extension Δx

2.4 Examples of simple harmonic motion

By definition, an object whose acceleration is proportional to the displacement from the equilibrium position, and which is always directed towards the equilibrium position, is undergoing simple harmonic motion. We now look at two simple examples of oscillatory motion which approximate to this definition. They may easily be set up as demonstrations of s.h.m.

Mass on a helical spring

Figure 2.19 illustrates a mass m suspended from a spring. (This sort of spring is called a *helical* spring because it has the shape of a helix.)

The weight mg of the mass is balanced by the tension T in the spring. When the spring is extended by an amount Δx, as in Figure 2.20, there is an additional upward force in the spring given by

$$F = -k\Delta x$$

where k is a constant for a particular spring, known as the spring constant. The spring constant is a measure of the stiffness of the spring. A stiff spring has a large value of k; a more flexible spring has a smaller value of k and, for the same force, would have a larger extension than one with a large spring constant. The spring constant k is given by the expression

$$k = \frac{F}{\Delta x}$$

and is measured in newtons per metre (N m^{-1}).

When the mass is released, the restoring force F pulls the mass towards the equilibrium position. (The minus sign in the expression for F shows the direction of this force.) The restoring force is proportional to the displacement. This means that the acceleration of the mass is proportional to the displacement from the equilibrium position and is directed towards the equilibrium position. This is the condition for simple harmonic motion.

The full theory shows that, for a mass m suspended from a light spring having spring constant k, the period T of the oscillations is given by

$$T = 2\pi\sqrt{\frac{m}{k}}$$

For oscillations to be simple harmonic, the spring must obey Hooke's law throughout – that is, the extensions must not exceed the limit of proportionality. Furthermore, for large amplitude oscillations, the spring may become slack. Ideally, the spring would have no mass, but if the suspended mass is more than about 20 times the mass of the spring, the error involved in assuming that the spring has no mass is less than 1%.

This example of simple harmonic motion is particularly useful in modelling the vibrations of molecules. A molecule containing two atoms oscillates as if the atoms were connected by a tiny spring. The spring constant of this spring depends upon the type of bonding between the atoms. The frequency of oscillation of the molecule can be measured experimentally using spectroscopy, and this gives direct information about the bonding. This model can be extended to solids, where atoms are often thought of as being connected to their neighbours by springs. Again, this leads to an experimental way of obtaining information about interatomic forces in the solid.

The simple pendulum

Figure 2.21 illustrates a simple pendulum. Ideally, a simple pendulum is a point mass m on a light, inelastic string. In real experiments we use a pendulum bob of finite size. When the bob is pulled aside through an angle θ and then released, there will be a restoring force acting in the direction of the equilibrium position.

Because the simple pendulum moves in the arc of a circle, the displacement will be an angular displacement θ rather than the linear displacement x we have been using so far.

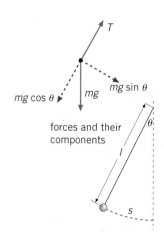

Figure 2.21 The simple pendulum

The two forces on the bob are its weight mg and the tension T in the string. The component of the weight along the direction of the string, $mg \cos \theta$, is equal to the tension in the string. The component of the weight at right angles to the direction of the string, $mg \sin \theta$, is the restoring force F. This makes the bob accelerate towards the equilibrium position.

The restoring force depends on $\sin \theta$. As θ increases, the restoring force is not proportional to the displacement (in this case θ), and so the motion is oscillatory but not simple harmonic. However, the situation is different if the angle θ is kept small (less than about 5°). For these small angles, θ is proportional to $\sin \theta$. In fact, if θ is measured in radians, then

θ in radians $\approx \sin \theta$

You can check this using your calculator. Some values of $\theta/°$, θ/rad and $\sin \theta$ are given in Table 2.1.

Table 2.1

$\theta/°$	θ/rad	$\sin \theta$
1.00	0.0175	0.0175
2.00	0.0349	0.0349
3.00	0.0524	0.0523
5.00	0.0873	0.0871
10.00	0.1745	0.1736

This means that, for small-amplitude oscillations (the angle of the string to the vertical should be less than about 5°), the pendulum bob oscillates with simple harmonic motion.

A full treatment of the theory shows that the period T is related to the length l of the pendulum (l is the distance between the centre of mass of the pendulum bob and the point of suspension – see Figure 2.21) by the expression

$$T = 2\pi \sqrt{\frac{l}{g}}$$

where g is the acceleration of free fall. An experiment in which the period of a simple pendulum is measured can be used to determine the acceleration of free fall. The experiment is repeated for different lengths of pendulum, and the gradient of a graph of T^2 against l is $4\pi^2/g$. This provides an alternative to dynamics experiments in which the time for a body to fall through measured distances is determined, and g is calculated from the equation of motion (see Chapter 2 of the AS book).

Example

A helical spring is clamped at one end and hangs vertically. It extends by 10 cm when a mass of 50 g is hung from its free end. Calculate:

(a) the spring constant of the spring,
(b) the period of small amplitude oscillations of the mass.

(a) Using $k = F/\Delta x$, the spring constant
$$k = 50 \times 10^{-3} \times 9.8/10 \times 10^{-2} = \textbf{4.9 N m}^{-1}$$
(b) Using $T = 2\pi\sqrt{(m/k)}$, the period
$$T = 2\pi\sqrt{(50 \times 10^{-3}/4.9)} = \textbf{0.63 s}$$

Now it's your turn

(a) The acceleration of free fall at the Earth's surface is 9.8 m s^{-2}. Calculate the length of a simple pendulum which would have a period of 1.0 s.
(b) The acceleration of free fall on the Moon's surface is 1.6 m s^{-2}. Calculate, for the pendulum in (a), its period of oscillation on the Moon.

Sections 2.2–2.4 Summary

- The period of an oscillation is the time taken to complete one oscillation.
- Frequency is the number of oscillations per unit time.
- Frequency f is related to period T by the expression: $f = 1/T$
- The displacement of a particle is its distance from the equilibrium position.
- Amplitude is the maximum displacement.
- Simple harmonic motion (s.h.m.) is defined as the motion of a particle about a fixed point such that its acceleration a is proportional to its displacement x from the fixed point, and is directed towards the fixed point: $a \propto -x$ or $a = -(2\pi f)^2 x$
- For a particle oscillating in s.h.m. with frequency f, then: $\omega = 2\pi f$
- The constant ω in the defining equation for simple harmonic motion is known as the angular frequency.
- Simple harmonic motion is described in terms of displacement x, amplitude A, frequency f by the following relations.

displacement:	$x = A \sin 2\pi f t$	or $x = A \cos 2\pi f t$
velocity:	$v = A 2\pi f \cos 2\pi f t$	or $v = -A 2\pi f \sin 2\pi f t$
acceleration:	$a = -A(2\pi f)^2 \sin 2\pi f t$	or $a = -A(2\pi f)^2 \cos 2\pi f t$

Sections 2.2–2.4 Questions

1 A particle is oscillating in simple harmonic motion with period 4.5 ms and amplitude 3.0 cm. At time $t = 0$, the particle is at the equilibrium position. Calculate, for this particle:
   (a) the frequency,
   (b) the angular frequency,
   (c) the maximum speed,
   (d) the magnitude of the maximum acceleration,
   (e) the speed at time $t = 1.0$ ms.

2 A particle is oscillating in simple harmonic motion with frequency 50 Hz and amplitude 15 mm. Calculate the speed when the displacement from the equilibrium position is 12 mm.

3 Geologists use the fact that the period of oscillation of a simple pendulum depends on the acceleration of free fall to map variations of g. A geologist determines the frequency of oscillation of a test pendulum of effective length 515.6 mm to be 0.6948 Hz. Calculate the acceleration of free fall at this locality.

4 A spring stretches by 85 mm when a mass of 50 g is hung from it. The spring is then stretched a further distance of 15 mm from the equilibrium position, and the mass is released at time $t = 0$. Calculate:
   (a) the spring constant,
   (b) the amplitude of the oscillations,
   (c) the period,
   (d) the displacement at time $t = 0.20$ s. $(g = 9.8\ \mathrm{m\ s^{-2}})$

2.5 Energy changes in simple harmonic motion

At the end of Section 2.5 you should be able to:
- describe and explain the interchange between kinetic and potential energy during simple harmonic motion.

Kinetic energy

In Section 2.3 we saw that the velocity of a particle vibrating with simple harmonic motion varies with time and, consequently, with the displacement of the particle. For the case where displacement x is zero at time $t = 0$, displacement and velocity are given by

$$x = A \sin 2\pi ft$$

and

$$v = A2\pi f \cos 2\pi ft$$

There is a trigonometrical relation between the sine and the cosine of an angle θ, which is $\sin^2 \theta + \cos^2 \theta = 1$. Applying this relation, we have

$$\frac{x^2}{A^2} + \frac{v^2}{A^2(2\pi f)^2} = 1$$

which leads to

$$v^2 = A^2(2\pi f)^2 - x^2(2\pi f)^2$$

and so

$$v = \pm 2\pi f \sqrt{A^2 - x^2}$$

(If we had taken the displacement and velocity equations for the case when the displacement is a maximum at $t = 0$, exactly the same relation would have been obtained. Try it!)

The kinetic energy of the particle (of mass m) oscillating with simple harmonic motion is $\frac{1}{2}mv^2$. Thus, the kinetic energy E_k at displacement x is given by

$$E_k = \frac{1}{2}m(2\pi f)^2(A^2 - x^2)$$

The variation with displacement of the kinetic energy is shown in Figure 2.22.

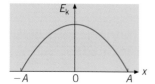

Figure 2.22 Variation of kinetic energy in s.h.m.

Potential energy

In Section 2.3 we stated that the defining equation for simple harmonic motion could be expressed in terms of the restoring force acting on the particle. Since $F = ma$ and $a = -(2\pi f)^2 x$ then at displacement x, this force is

$$F_{\text{res}} = -m(2\pi f)^2 x$$

where m is the mass of the particle. To find the change in potential energy of the particle when the displacement increases by Δx, we need to find the work done against the restoring force.

The work done in moving the point of application of a force F by a distance Δx is $F\Delta x$. In the case of the particle undergoing simple harmonic motion, we know that the restoring force is directly proportional to displacement. To calculate the work done against the restoring force in giving the particle a displacement x, we take account of the fact that F_{res} depends on x by taking the average value of F_{res} during this displacement. The average value is just $\frac{1}{2}m(2\pi f)^2 x$, since the value of F_{res} is zero at $x = 0$ and increases linearly to $m(2\pi f)^2 x$ at displacement x. Thus, the potential energy E_p at displacement x is given by *average restoring force × displacement*, or

$$E_p = \frac{1}{2}m(2\pi f)^2 x^2$$

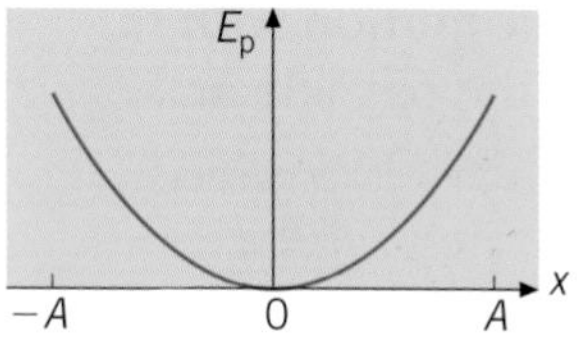

Figure 2.23 Variation of potential energy in s.h.m.

The variation with displacement of the potential energy is shown in Figure 2.23.

Total energy

The total energy E_{tot} of the oscillating particle is given by

$$
\begin{aligned}
E_{tot} &= E_k + E_p \\
&= \tfrac{1}{2} m(2\pi f)^2(A^2 - x^2) + \tfrac{1}{2} m(2\pi f)^2 x^2 \\
E_{tot} &= \tfrac{1}{2} m(2\pi f)^2 A^2
\end{aligned}
$$

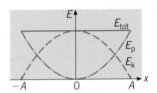

Figure 2.24 Energy variations in s.h.m.

This total energy is constant since m, $(2\pi f)$ and A are all constant. We might have expected this result, as it merely expresses the law of conservation of energy.

The variations with displacement x of the total energy E_{tot}, the kinetic energy E_k and the potential energy E_p are shown in Figure 2.24.

Example

A particle of mass 60 g oscillates in simple harmonic motion with frequency 1.0 Hz and amplitude 15 mm. Calculate:
(a) the total energy,
(b) the kinetic and potential energies at half-amplitude
    (at displacement $x = 7.5$ mm).

(a) Using $E_{tot} = \tfrac{1}{2} m(2\pi f)^2 A^2$,

$$
\begin{aligned}
E_{tot} &= \tfrac{1}{2} \times 60 \times 10^{-3} \times (2\pi)^2 \times (15 \times 10^{-3})^2 \\
&= \mathbf{2.7 \times 10^{-4}\,J}
\end{aligned}
$$

(don't forget to convert g to kg and mm to m)

(b) Using $E_k = \tfrac{1}{2} m(2\pi f)^2 (A^2 - x^2)$,

$$
\begin{aligned}
E_k &= \tfrac{1}{2} \times 60 \times 10^{-3} \times (2\pi)^2 \times [(15 \times 10^{-3})^2 - (7.5 - 10^{-3})^2] \\
&= \mathbf{2.0 \times 10^{-4}\,J}
\end{aligned}
$$

Using $E_p = \tfrac{1}{2} m(2\pi f)^2 x^2$,

$$
\begin{aligned}
E_p &= \tfrac{1}{2} \times 60 \times 10^{-3} \times (2\pi)^2 \times (7.5 \times 10^{-3})^2 \\
&= \mathbf{0.7 \times 10^{-4}\,J}
\end{aligned}
$$

Note that $E_{tot} = E_k + E_p$, as expected.

Now it's your turn

A particle of mass 0.40 kg oscillates in simple harmonic motion with frequency 5.0 Hz and amplitude 12 cm. Calculate, for the particle at displacement 10 cm:
(a) the kinetic energy,
(b) the potential energy,
(c) the total energy.

Section 2.5 Summary

- The kinetic energy E_k of a particle of mass m oscillating in simple harmonic motion with frequency f and amplitude A is:
$$E_k = \tfrac{1}{2}m(2\pi f)^2(A^2 - x^2)$$
where x is the displacement.

- The potential energy E_p of a particle of mass m oscillating in simple harmonic motion with frequency f is:
$$E_p = \tfrac{1}{2}m(2\pi f)^2 x^2$$
where x is the displacement.

- The total energy E_{tot} of a particle of mass m oscillating in simple harmonic motion with frequency f and amplitude A is:
$$E_{tot} = \tfrac{1}{2}m(2\pi f)^2 A^2$$

- For a particle oscillating in simple harmonic motion:
$$E_{tot} = E_k + E_p$$
This expresses the law of conservation of energy.

Section 2.5 Questions

1 One particle oscillating in simple harmonic motion has ten times the total energy of another particle, but the frequencies and masses are the same. Calculate the ratio of the amplitudes of the two motions.

2 **(a)** Calculate the displacement, expressed as a fraction of the amplitude A, of a particle moving in simple harmonic motion with a speed equal to half the maximum value.

(b) Calculate the displacement at which the energy of the particle has equal amounts of kinetic and of potential energy.

2.6 Free and damped oscillations

At the end of Section 2.6 you should be able to:
- describe the effects of damping on an oscillatory system
- describe practical examples of forced oscillations and resonance
- describe graphically how the amplitude of a forced oscillation changes with frequency near to the natural frequency of the system
- describe examples where resonance is useful and other examples where resonance should be avoided.

A particle is said to be undergoing **free oscillations** when the only external force acting on it is the restoring force.

There are no forces to dissipate energy and so the oscillations have constant amplitude. Total energy remains constant. This is the situation we have been considering so far. Simple harmonic oscillations are free oscillations.

In real situations, however, frictional and other resistive forces cause the oscillator's energy to be dissipated, and this energy is converted eventually into heat energy. The oscillations are said to be **damped**.

The total energy of the oscillator decreases with time. The damping is said to be light when the amplitude of the oscillations decreases gradually with time. This is illustrated in Figure 2.25. The decrease in amplitude is, in fact, exponential with time. The period of the oscillation is slightly greater than that of the corresponding free oscillation.

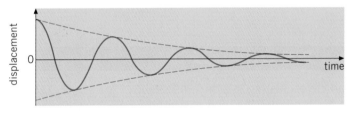

Figure 2.25 Lightly damped oscillations

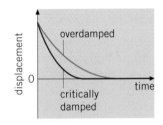

Figure 2.26 Critical damping and overdamping

Figure 2.27 Vehicle suspension system showing springs and dampers

Heavy damping causes the oscillations to die away more quickly. If the damping is increased further, then the system reaches **critical damping** point. Here the displacement decreases to zero in the shortest time, without any oscillation (Figure 2.26).

Any further increase in damping produces **overdamping**. The displacement decreases to zero in a longer time than for critical damping (Figure 2.26).

Damping is often useful in an oscillating system. For example, vehicles have springs between the wheels and the frame to give a smoother and more comfortable ride (Figure 2.27). If there was no damping, a vehicle would move up and down for some time after hitting a bump in the road. Dampers (shock absorbers) are connected in parallel with the springs so that the suspension has critical damping and comes to rest in the shortest time possible. Dampers often work through hydraulic action. When the spring is compressed, a piston connected to the vehicle frame forces oil through a small hole in the piston, so that the energy of the oscillation is dissipated as heat energy in the oil.

Many swing doors have a damping mechanism fitted to them. The purpose of the damper is so that the open door, when released, does not overshoot the closed position with the possibility of injuring someone approaching the door. Most door dampers operate in the overdamped mode.

Forced oscillations and resonance

When a vibrating body undergoes free (undamped) oscillations, it vibrates at its **natural frequency**. We have already met the idea of a natural frequency in Chapter 6 of the AS book, when talking about stationary waves on strings. The natural frequency of such a system is the frequency of the first mode of vibration; that is, the fundamental frequency. A practical example is a guitar string, plucked at its centre, which oscillates at a particular frequency which depends on the speed of progressive waves on the string and the length of the string. The speed of progressive waves on the string depends on the mass per unit length of the string and the tension in the string.

Vibrating objects may have periodic forces acting on them. These periodic forces will make the object vibrate at the frequency of the applied force, rather than at the natural frequency of the system. The object is then said to be undergoing **forced vibrations**. Figure 2.28 illustrates apparatus which may be used to demonstrate the forced vibrations of a mass on a helical spring. The vibrator provides the forcing (driving) frequency.

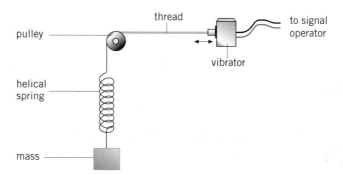

Figure 2.28 Demonstration of forced oscillations

As the frequency of the vibrator is gradually increased from zero, the mass begins to oscillate. At first the amplitude of the oscillations is small, but it increases with increasing frequency. When the driving frequency equals the natural frequency of oscillation of the mass–spring system, the amplitude of the oscillations reaches a maximum. The frequency at which this occurs is called the **resonant frequency**, and **resonance** is said to occur. If the driving frequency is increased further, the amplitude of oscillation of the mass decreases. The variation with driving frequency f of the amplitude A of vibration of the mass is illustrated in Figure 2.29. This graph is often called a **resonance curve**.

> Resonance occurs when the natural frequency of vibration of an object is equal to the driving frequency, giving a maximum amplitude of vibration.

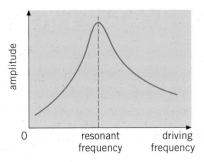

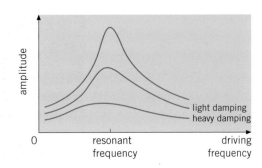

Figure 2.29 Resonance curve

Figure 2.30 Effect of damping on the resonance curve

The effect of damping on the amplitude of forced oscillations can be investigated by attaching a light but stiff card to the mass in Figure 2.28. Movement of the card gives rise to air resistance and thus damping of the oscillations. The degree of damping may be varied by changing the area of the card. The effects of damping are illustrated in Figure 2.30. It can be seen that, as the degree of damping increases:

- the amplitude of oscillation at all frequencies is reduced
- the frequency at maximum amplitude shifts gradually towards lower frequencies
- the peak becomes flatter.

Barton's pendulums may also be used to demonstrate resonance and the effects of damping. The apparatus consists of a set of light pendulums, made (for example) from paper cones, and a more massive pendulum (the driver), all supported on a taut string. The arrangement is illustrated in Figure 2.31. The lighter pendulums have different lengths, but one has the same length as the driver. This has the same natural frequency as the driver and will, therefore, vibrate with the largest frequency of all the pendulums (Figure 2.32).

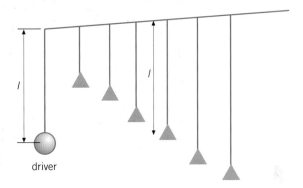

driver

Figure 2.31 Barton's pendulums

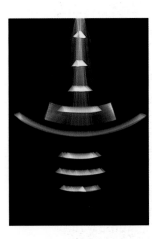

Figure 2.32 Time-exposure photographs of Barton's pendulums with light damping, taken end-on. The longest arc, in the middle, is the driver.

Adding weights to the paper cones reduces the effect of damping. With less damping, the amplitude of the resonant pendulum is much larger.

There are many examples of resonance in everyday life. One of the simplest is that of pushing a child on a swing (Figure 2.33). We push at the same frequency as the natural frequency of oscillation of the swing and child, so that the amplitude of the motion increases.

Figure 2.33 Pushing a child on a swing makes the swing go higher

The operation of the engine of a vehicle causes a periodic force on the parts of the vehicle, which can cause them to resonate. For example, at particular frequencies of rotation of the engine, the mirrors may resonate. To prevent excessive vibration, the mountings of the mirrors provide damping.

Musical instruments rely on resonance to amplify the sound produced. The sound from a tuning fork is louder when it is held over a tube of just the right length, so that the column of air resonates. We have met this phenomenon in Chapter 6 of the AS book, in connection with the resonance tube method of measuring the speed of sound in air. Stringed instruments have a hollow wooden box with a hole under the strings which acts in a similar way. To amplify all notes from all of the strings, the sounding-box has to be a complex shape so that it resonates at many different frequencies.

A spectacular example of resonance that is often quoted is the failure of the first suspension bridge over the Tacoma Narrows in Washington State, USA.

Figure 2.34 The Tacoma Narrows bridge disaster

Wind caused the bridge to oscillate. It was used for months even though the roadway was oscillating with transverse vibrations. Approaching vehicles would appear, and then disappear, as the bridge deck vibrated up and down. One day strong winds set up twisting vibrations (Figure 2.34) and the amplitude of vibration increased due to resonance, until eventually the bridge collapsed. The driver of a car that was on the bridge managed to walk to safety before the collapse, although his dog could not be persuaded to leave the car!

Section 2.6 Summary

- Free oscillations are oscillations where there are no resistive forces acting on the oscillating system.
- Damping is produced by resistive forces which dissipate the energy of the vibrating system.
- Light damping causes the amplitude of vibration of the oscillation to decrease gradually. Critical damping causes the displacement to be reduced to zero in the shortest time possible, without any oscillation of the object. Overdamping also causes an exponential reduction in displacement, but over a greater time than for critical damping.
- The natural frequency of vibration of an object is the frequency at which the object will vibrate when allowed to do so freely.
- Forced oscillations occur when a periodic driving force is applied to a system which is capable of vibration.
- Resonance occurs when the driving frequency on the system is equal to its natural frequency of vibration. The amplitude of vibration is a maximum at the resonant frequency.

Section 2.6 Question

The apparatus of Figure 2.28 is used to demonstrate forced vibrations and resonance. A 50 g mass is suspended from the spring, which has a spring constant of $7.9 \, \text{N m}^{-1}$.

(a) Calculate the resonant frequency f_0 of the system.

(b) A student suggests that resonance should also be observed at a frequency of $2f_0$. Discuss this suggestion.

Exam-style Questions

1 All bodies moving in a circle experience centripetal acceleration.
   (a) What does *centripetal* mean? Explain how the centripetal acceleration arises.
   (b) The Moon's orbit around the Earth may be assumed to be circular, with radius 3.84×10^5 km. The Moon moves with constant speed in the orbit, and takes 27.3 days to complete one orbit. Calculate the magnitude of the centripetal acceleration of the Moon.

2 A bicycle is set up on a test rig with its front wheel clamped and its rear wheel resting on a roller. The pedals are turned so that, if the bicycle had been free, it would have been travelling at a speed of 8.3 m s^{-1}. The outside diameter of each tyre is 0.66 m.
   (a) Calculate:
      (i) the angular velocity of the rear wheel of the bicycle,
      (ii) how many times per second the rear wheel rotates.
   (b) A small pebble of mass 4.0 g is embedded in the tread of the rear tyre.
      (i) Calculate the magnitude, and state the direction, of the force needed to keep the pebble in circular motion.
      (ii) This force is equal to the maximum frictional force that the tyre can exert on the pebble. Describe and explain what happens when the rear wheel is rotated at a faster rate.

3 **(a)** Define simple harmonic motion.
   (b) Figure 2.35a illustrates a U-tube of uniform cross-sectional area A containing liquid of density ρ. The total length of the liquid column is L. When the liquid is displaced by an amount Δx from its equilibrium position (see Figure 2.35b), it oscillates with simple harmonic motion.
      (i) The weight of liquid above AB in Figure 2.35b provides the restoring force. Write down an expression for the restoring force.
      (ii) Write down an expression for the acceleration of the liquid column caused by this force.
      (iii) Explain how this fulfils the condition for simple harmonic motion.
      (iv) Write down an expression for the frequency of the oscillations.

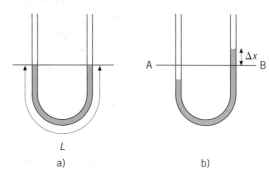

Figure 2.35

4 A 'baby bouncer' consists of a harness attached to a rubber cord. A baby of mass 6.5 kg is placed gently in the harness and the cord extends by 0.40 m. The baby is then pulled down another 0.10 m and, when released, begins to move with simple harmonic motion. Calculate:
   (a) the period of the motion,
   (b) the maximum force on the baby.
   (acceleration of free fall $g = 9.8$ m s^{-2})

5 A simple model of a hydrogen molecule assumes that it consists of two oscillating hydrogen atoms connected by a spring of spring constant 1.1×10^3 N m^{-1}.
   (a) The mass of a hydrogen atom is 1.7×10^{-27} kg. Calculate the frequency of oscillation of the hydrogen molecule.
   (b) Explain why light of wavelength $2.3\,\mu$m would be strongly absorbed by this model of the hydrogen molecule.
   (speed of light $c = 3.0 \times 10^8$ m s^{-1})

6 If you walk holding a full cup of coffee, you may find that, for a particular cup size and rate of taking steps, the coffee sloshes above the side of the cup, whereas if you walk more slowly, there is no risk of spillage. Explain this phenomenon, and estimate the critical combination of cup diameter and rate of taking steps.

7 As a party trick, the operatic tenor Enrico Caruso used to shatter a wine glass by singing a note of just the right frequency at full volume. If Caruso had been a physicist, what would he have understood by 'just the right frequency'? Why might he not have been successful if the glass had still contained some wine?

3. Gravitational fields

This chapter explores the nature of gravitational fields. Gravitational forces are responsible for the birth of stars in giant clouds in space. Gravity also helps us to keep our feet on the ground! We shall look at gravitational field lines and strength and how this relates to mass. Gravity is also involved in the way in which objects orbit one another, from planets orbiting the Sun to the geostationary satellites around our own world.

3.1 Gravitational fields

At the end of Section 3.1 you should be able to:
- state Newton's law of gravitation
- select and use the equation $F = -GMm/r^2$ for the force between two point masses or spherical objects
- use gravitational field lines to represent a gravitational field
- define *gravitational field strength* as force per unit mass
- select and apply the equation $g = -GM/r^2$ for the gravitational field strength of a point mass
- select and use the equation $g = -GM/r^2$ to determine the mass of the Earth or another similar object
- explain that, close to the Earth's surface, the gravitational field strength is uniform and approximately equal to the acceleration of free fall.

Force between masses

We are familiar with the fact that the Earth's force of gravity is responsible for our weight, the force which pulls us towards the Earth. Isaac Newton concluded that the Earth's force of gravity is also responsible for keeping the Moon in orbit. His calculations showed that the Earth's force of gravity extends into space, but weakens with distance according to an inverse square law. That is, the Earth's force of gravity varies inversely with the square of the distance from the centre of the Earth. If you go two Earth radii from the Earth's surface, the force is a quarter of the force on the Earth's surface. This is part of **Newton's law of gravitation**.

Newton's law of gravitation states that two point masses attract each other with a force that is proportional to the product of their masses and inversely proportional to the square of their separation.

Hence, if F is the force of attraction between two bodies of mass M and m respectively with distance r between their centres, then

$F \propto Mm/r^2$

or

$$F = -\frac{GMm}{r_2}$$

where the constant of proportionality G is called the gravitational constant. The minus sign indicates that as r increases, F decreases.

The value of G is 6.67×10^{-11} N m^{2}kg^{-2}.

You will notice that this equation has a similar form to that for Coulomb's law between two charges (page 84). Both Newton's law and Coulomb's law are inverse square laws of force. We say that the equations are analagous, or that there is an analogy between this aspect of electric fields and gravitational fields. However, there are important differences.

■ The electric force acts on charges, whereas the gravitational force acts on masses.
■ The electric force can be attractive or repulsive, depending on the signs of the interacting charges, whereas two masses always attract each other.

It is possible to measure the gravitational constant G in a school laboratory, but the force of gravity between laboratory-sized masses is so small that it is not easy to obtain a reliable result.

Gravitational field lines

A region of space where a small mass experiences a force is known as a gravitational field. The direction of this gravitational field may be represented by gravitational field lines.

A gravitational field line is the direction in which a small test mass would move when placed in the field.

The gravitational field around a uniform spherical mass is shown in Figure 3.1.

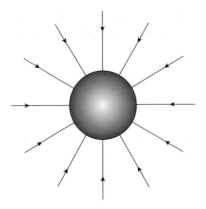

Figure 3.1 Gravitational field around a uniform spherical mass

Note that, since gravitational forces are always attractive, the field is directed towards the mass.

All the field lines appear to converge to the centre of the sphere. As a result, the total mass of a uniform sphere may be considered as a point mass at the centre of the sphere.

Gravitational field strength

> The **gravitational field strength** at a point is defined as the force per unit mass acting on a small mass placed at that point.

The force per unit mass is also a measure of the acceleration of free fall, by Newton's second law of motion $F = ma$ (See Chapter 1). Thus, the gravitational field strength is given by $F/m = g$, where F is in newtons and m is in kilograms. This means that the gravitational field strength at the Earth's surface is about $9.8\,\mathrm{N\,kg^{-1}}$. The unit $\mathrm{N\,kg^{-1}}$ is equivalent to the unit of acceleration, $\mathrm{m\,s^{-2}}$.

Field strength due to an isolated point mass

We have already met the idea of a point mass in talking about the motion of a particle (Chapter 2 of the AS book). Of course, all the masses we come across in the laboratory have a finite size. But for calculations involving gravitational forces, it is fortunate that a spherical mass behaves as if it were a point mass at the centre of the sphere, with all the mass of the sphere concentrated at that point. This is very similar to the idea that the charge on a conducting sphere can be considered to be concentrated at the centre of the sphere.

The gravitational field outside a spherical mass is a radial one, just as the field associated with a point charge is radial. From Newton's law, the attractive force on a mass m caused by another mass M, with a distance of r between their centres, is given by

$$F = -\frac{GMm}{r^2}$$

This means that the force per unit mass or gravitational field strength g is given by

$$g = \frac{F}{m} = -\frac{GM}{r^2}$$

The field strengths due to masses that you find in a laboratory are tiny. For example, the field strength one metre away from an isolated mass of one kilogram is only $7 \times 10^{-11} \, \text{N kg}^{-1}$. But field strengths due to the masses of objects such as the Earth or Moon are much larger. We already know that the field strength due to the Earth at the surface of the Earth is about $10 \, \text{N kg}^{-1}$. We can use this information to deduce information about the Earth, for example the mass of the Earth. Look at the Example that follows.

Example

The radius of the Earth is $6.4 \times 10^6 \, \text{m}$ and the gravitational field strength at its surface is $9.8 \, \text{N kg}^{-1}$.
(a) Assuming that the field is radial, calculate the mass of the Earth.
(b) The radius of the Moon's orbit about the Earth is $3.8 \times 10^8 \, \text{m}$. Calculate the strength of the Earth's gravitational field at this distance.
(c) The mass of the Moon is $7.4 \times 10^{22} \, \text{kg}$. Calculate the gravitational attraction between the Earth and the Moon.
(gravitational constant $G = 6.7 \times 10^{-11} \, \text{N m}^2 \text{kg}^{-2}$)

(a) Using $g = GM/r^2$, we have
$M = gr^2/G = 9.8 \times (6.4 \times 10^6)^2/6.7 \times 10^{-11} = \textbf{6.0} \times \textbf{10}^{\textbf{24}} \, \textbf{kg}$
(b) Using $g = GM/r^2$, we have
$g = 6.7 \times 10^{-11} \times 6.0 \times 10^{24}/(3.8 \times 10^8)^2 = \textbf{2.8} \times \textbf{10}^{\textbf{-3}} \textbf{N kg}^{\textbf{-1}}$
(c) Using $F = GMm/r^2$, we have
$F = 6.7 \times 10^{-11} \times 6.0 \times 10^{24} \times 7.4 \times 10^{22}/(3.8 \times 10^8)^2 = \textbf{2.1} \times \textbf{10}^{\textbf{20}} \textbf{N}$

Now it's your turn

The mass of Jupiter is $1.9 \times 10^{27} \, \text{kg}$ and its radius is $7.1 \times 10^7 \, \text{m}$. Calculate the gravitational field strength at the surface of Jupiter.
(gravitational constant $G = 6.7 \times 10^{-11} \, \text{N m}^2 \text{kg}^{-2}$)

Field strength at the Earth's surface

The gravitational field strength at the surface of the Earth, which is equal to the acceleration of free fall g, is given by the expression

$$g = -GM/R^2$$

where M is the mass of the Earth (6.0×10^{24} kg) and R is its radius (6.4×10^6 m).

This radius R is large when compared with changes in altitude that we normally experience on the Earth's surface. For example, Mount Everest has a height of 8.8×10^3 m. So, if the distance from the centre of the Earth changes by h, where h is much less than the radius R, then

$$R^2 \approx (R \pm h)^2$$

This means that, near the surface of the Earth, the gravitational field strength is constant (uniform).

This can be understood by reference to Figure 3.1. The Earth has a large radius and therefore, any small area on its surface will be approximately plane. The gravitational field lines are normal to this plane surface and will, therefore be parallel, indicating a constant (uniform) field.

Mass and weight

Mass is said to be a measure of the inertia of a body to changes in velocity. Unless the body is travelling at speeds close to that of light, its mass is constant.

In a gravitational field, by definition, there is a force acting on the mass equal to the product of mass and gravitational field strength. This force is called the **weight**.

For an object of mass m in a gravitational field of strength g, the weight W is given by

weight = mass × gravitational field strength

or

$$W = mg$$

Although mass is invariant, weight depends on gravitational field strength. For example, a person of mass 60 kg has a weight of 600 N on Earth, but only 100 N on the Moon, although the mass is still 60 kg.

3.2 Gravity and orbits

At the end of Section 3.2 you should be able to:

- analyse circular orbits in an inverse square law field by relating the gravitational force to the centripetal acceleration it causes
- define and use the *period* of an object describing a circle
- derive the equation $T^2 = 4\pi^2 r^3/GM$ from first principles
- select and apply the equation $T^2 = 4\pi^2 r^3/GM$ for planets and satellites (natural and artificial)
- select and apply Kepler's third law $T^2 \propto r^3$ to solve problems
- define *geostationary orbit* of a satellite and state the uses of such satellites.

Planet and satellite orbits

Most planets in the solar system have orbits which are nearly circular. We now bring together the idea of a gravitational force and that of a centripetal force (see Chapter 2) to derive a relation between the period and the radius of the orbit of a planet describing a circular path about the Sun, or a satellite moving round the Earth or another planet.

Consider a planet of mass m in circular orbit about the Sun, of mass M. If the radius of the orbit is r, the gravitational force F_{grav} between Sun and planet is

$$F_{grav} = GMm/r^2$$

by Newton's law of gravitation. It is this force that provides the centripetal force as the planet moves in its orbit. The centripetal force F_{circ} is given by

$$F_{circ} = mv^2/r$$

where v is the linear speed of the planet. As has just been stated,

$$F_{grav} = F_{circ}$$

Thus

$$GMm/r^2 = mv^2/r$$

The period T of the planet in its orbit is the time required for the planet to travel a distance $2\pi r$. It is moving at speed v, so

$$v = 2\pi r/T$$

Putting this into the equation above, we have

$$GMm/r^2 = m(4\pi^2 r^2/T^2)/r$$

or, simplifying,

$$T^2 = (4\pi^2/GM)r^3$$

Another way of writing this is

$$\frac{T^2}{r^3} = \frac{4\pi^2}{GM}$$

Look at the right-hand side of this equation. The quantities π and G are constants. If we are considering the relation between T and r for planets in the solar system, then M is the same for each planet because it is the mass of the Sun.

This equation shows that:

> For planets or satellites describing circular orbits about the same central body, the square of the period is proportional to the cube of the radius of the orbit.

This relation is known as **Kepler's third law of planetary motion**. Johannes Kepler (1571–1630) analysed data collected by Tycho Brahe (1546–1601) on planetary observations. He showed that the observations fitted a law of the form $T^2 \propto r^3$. Fifty years later Newton showed that an inverse square law of gravitation, the idea of centripetal acceleration, and the second law of motion, gave an expression of exactly the same form. Newton cited Kepler's law in support of his law of gravitation. In fact, the orbits of the planets are not circular, but elliptical, a fact recognised by Kepler. The derivation is simpler for the case of a circular orbit.

Table 3.1 collects information about T and r for planets of the solar system. The last column shows that the value of T^2/r^3 is indeed a constant. Moreover, the value of T^2/r^3 agrees very well with the value of $4\pi^2/GM$, 2.97×10^{-25} yr^2 m^{-3}.

Table 3.1

planet	T/(Earth years)	r/km	T^2/r^3(yr^2 m^{-3})
Mercury	0.241	57.9×10^6	2.99×10^{-25}
Venus	0.615	108.0×10^6	3.00×10^{-25}
Earth	1.00	150.0×10^6	2.96×10^{-25}
Mars	1.88	228.0×10^6	2.98×10^{-25}
Jupiter	11.9	778.0×10^6	3.01×10^{-25}
Saturn	29.5	1430.0×10^6	2.98×10^{-25}
Uranus	84.0	2870.0×10^6	2.98×10^{-25}
Neptune	165	4500.0×10^6	2.99×10^{-25}
Pluto	248	5900.0×10^6	2.99×10^{-25}

(average 2.99×10^{-25})

Satellites are widely used in telecommunication. Many communication satellites are placed in what is called **geostationary orbit**. That is, they are in equatorial orbits with exactly the same period of rotation as the Earth, and move in the same direction as the Earth, so that they are always above the same point on the equator. Details of the orbit of such a satellite are worked out in the example which follows.

Example

For a geostationary satellite, calculate:
(a) the height above the Earth's surface,
(b) the speed in orbit.
(radius of Earth = 6.38×10^6 m; mass of Earth = 5.98×10^{24} kg)

(a) The period of the satellite is 24 hours = 8.64×10^4 s.
Using $T^2/r^3 = 4\pi^2/GM$, we have
$r^3 = 6.67 \times 10^{-11} \times 5.98 \times 10^{24} \times (8.64 \times 10^4)^2/4\pi^2$, giving
$r^3 = 7.54 \times 10^{22}$ m^3. Taking the cube root, the radius r of the orbit is
4.23×10^7 m. The distance above the Earth's surface is
$(4.23 \times 10^7 - 6.4 \times 10^6) =$ **3.59×10^7 m**.
(b) Since $v = 2\pi r/T$, the speed is given by
$v = 2\pi \times 4.23 \times 10^7/8.64 \times 10^4 =$ **3070 m s^{-1}**.

Now it's your turn

The radius of the Moon's orbit is 3.84×10^8 m, and its period is 27.4 days. Use Kepler's law to calculate the period of the orbit of a satellite orbiting the Earth just above the Earth's surface.
(radius of Earth = 6.38×10^6 m)

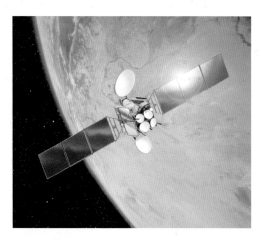

Figure 3.2 Communication satellite

Weightlessness

Suppose that you are carrying out an experiment involving the use of a newton balance in a lift. When the lift is stationary, an object of mass 10 kg, suspended from the balance, will give a weight reading of $10g$ N. If the lift accelerates upwards with acceleration $0.1g$, the reading on the balance increases to $11g$ N.

If the lift accelerates downwards with acceleration 0.1g, the apparent weight of the object decreases to 9g N. If, by an unfortunate accident, the lift cable were to break and there were no safety restraints, the lift would accelerate downwards with acceleration g. The reading on the Newton balance would be zero. If during the fall you were to drop the pencil with which you are recording the balance readings, it would not fall to the floor of the lift, but remain stationary with respect to you. Both you and the pencil are in free fall. You are experiencing **weightlessness**. Figure 3.3 illustrates your predicament. It might be more correct to refer to this situation as *apparent* weightlessness, as you can only be truly weightless in the absence of a gravitational field – that is, at an infinite distance from the Earth or any other attracting object.

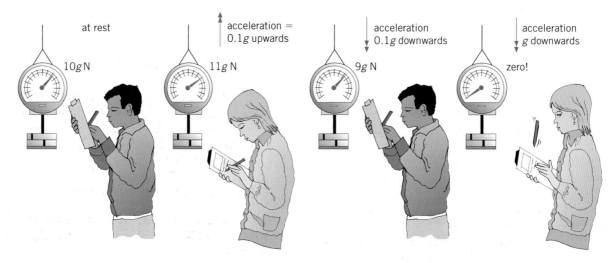

at rest   10g N   acceleration = 0.1g upwards   11g N   acceleration 0.1g downwards   9g N   acceleration g downwards   zero!

Figure 3.3 Lift experiment on weightlessness

A similar situation arises in a satellite orbiting the Earth. The force of gravity, which provides the centripetal force, is causing the body to fall out of its expected straight-line path. People and objects inside the satellite are experiencing a free fall situation and apparent weightlessness.

Sections 3.1–3.2 Summary

■ The attractive force between two point masses is proportional to the product of the masses and inversely proportional to the square of the distance between them. This is Newton's law of gravitation:
$F = -GMm/r^2$

■ A gravitational field is a region round a mass where another mass feels a force. Gravitational field strength g is the force per unit mass:
$g = F/m$
g is also the acceleration of free fall.

■ The gravitational field strength at a point in the gravitational field of a point mass M is: $g = -GM/r^2$

■ For a planet or satellite in circular orbit about a body, the square of the period is proportional to the cube of the radius of the orbit. This is Kepler's third law of planetary motion.

Sections 3.1–3.2 Questions

1   The Earth's radius is about 6.4×10^6 m. The acceleration of free fall at the Earth's surface is 9.8 m s^{-2}. The gravitational constant G is 6.7×10^{-11} N m^2 kg^{-2}. Use this information to estimate the mean density of the Earth.

2   The times for Mars and Jupiter to orbit the Sun are 687 days and 4330 days respectively. The radius of the orbit of Mars is 228×10^6 km. Calculate the radius of the orbit of Jupiter.

3   The weight of a passenger in an aircraft on the runway is W. His weight when the aircraft is flying at an altitude of 16 km above the Earth's surface is W_a. The percentage change F in his weight is given by $F = 100(W_a - W)/W\%$. Taking the radius of the Earth as 6.378×10^3 km, calculate F. Calculate also the percentage change P in his gravitational potential energy.

Exam-style Questions

1   Figure 3.4 illustrates the apparatus used by Cavendish in 1798 to find a value for the gravitational constant G. In a school experiment using similar apparatus, two lead spheres are attached to a light horizontal beam which is suspended by a wire. When a flask of mercury is brought close to each sphere, the gravitational attraction causes the beam to twist through a small angle. From measurements of the twisting (torsional) oscillations of the beam, a value can be found for the force producing a measured deflection. G can then be calculated if the large and small masses are known.

(a) In such an experiment, one lead sphere has mass 6.22×10^{-3} kg and the mass of the mercury flask is 0.713 kg. Calculate the force between them when they are 72.0 mm apart. (gravitational constant $G = 6.67 \times 10^{-11}$ N m^2 kg^{-2})

(b) Comment on the size of the force.

2   The gravitational field strength at the surface of the Moon is 1.62 N kg^{-1}. The radius of the Moon is 1740 km.

(a) Show that the mass of the Moon is 7.35×10^{22} kg.

(b) The Moon rotates about its axis (as well as moving in orbit about the Earth). In the future, scientists may wish to put a satellite into an orbit about the Moon such that the satellite remains stationary above one point on the Moon's surface.

   (i) Explain why this orbit must be an equatorial orbit.

   (ii) The period of rotation of the Moon about its axis is 27.4 days. Calculate the radius of the required orbit. ($G = 6.67 \times 10^{-11}$ N m^2 kg^{-2})

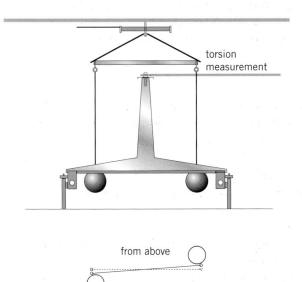

torsion measurement

from above

Figure 3.4 Cavendish's experiment for G

4. Thermal physics

The aim of this chapter is to describe a number of phenomena associated with temperature and thermal energy. We shall start by looking at what is meant by temperature and then see how scales of temperature are defined. The fundamental scale of temperature, the thermodynamic scale, is based on the properties of an ideal gas. To develop the idea of the temperature of a gas, we shall need to think of the gas as a collection of very small particles in continuous random motion. This is the basis of the kinetic theory of gases. Finally, we shall consider the energy of the particles of a gas and link this to internal energy.

4.1 Temperature

At the end of Section 4.1 you should be able to:

- explain that thermal energy is transferred from a region of higher temperature to a region of lower temperature
- explain that regions of equal temperature are in thermal equilibrium
- describe how there is an absolute scale of temperature which does not depend on the property of any particular substance (i.e. the thermodynamic scale), and the concept of absolute zero
- convert temperatures measured in kelvin to degrees Celsius (or vice versa) $T/K = \theta/^{\circ}C + 273.15$
- state that absolute zero is the temperature at which a substance has minimum internal energy.

Our everyday idea of temperature is based on our sense of touch. Putting your hand into a bowl containing ice immediately gives a sense of cold; putting the other hand into a bowl of warm water gives the sensation of heat (Figure 4.1). Intuitively, we would say that the water is at a higher temperature than the ice.

Think about a thick metal bar with one end at a higher temperature than the other. For example, one end can be heated by pouring hot water over it, and the other end cooled by holding it under a cold-water tap. The effect of the temperature difference between the ends is that thermal energy is transferred along the bar from the high temperature end to the low temperature end. We can think of this in terms of the vibrations of the atoms of the metal. One atom passes on some of its vibrational energy to its neighbour, which originally had less. If the bar is removed from the arrangement for keeping the ends at different temperatures, eventually the

Figure 4.1

whole of the bar will end up in equilibrium at the same temperature. When different regions in thermal contact are at the same temperature, they are said to be in **thermal equilibrium**.

In physics, we look for ways of defining and measuring quantities. In the case of temperature, we will first look at ways of measuring this quantity, and then think about the definition.

Many physical properties change with temperature. Most materials (solids, liquids and gases) expand as their temperature is increased. The electrical resistance of a metal wire increases as the temperature of the wire is increased. If two wires of different metals are twisted together at one end, and the other ends are connected to a voltmeter, the reading on the voltmeter depends on the temperature of the junction of the wires. All these properties may be made use of in different types of thermometer.

A thermometer is an instrument for measuring temperature. The physical property on which a particular thermometer is based is called the thermometric property, and the working material of the thermometer, the property of which varies with temperature, is called the thermometric substance. Thus, in the familiar mercury-in-glass thermometer, the thermometric substance is mercury and the thermometric property is the length of the mercury thread in the capillary tube of the thermometer.

Remember that temperature measures the degree of 'hotness' of a body. It does not measure the *amount* of heat energy.

Temperature scales

Each type of thermometer can be used to establish its own temperature scale. To do this, the fact that substances change state (from solid to liquid, or from liquid to gas) at fixed temperatures is used to define reference temperatures, which are called fixed points. By taking the value of the thermometric property at two fixed points, and dividing the range of values into a number of equal steps (or degrees), we can set up what is called an empirical scale of temperature for that thermometer. ('Empirical' means 'derived by experiment'.) If the fixed points are the melting point of ice and the boiling point of water, and if we choose to have one hundred equal degrees between the temperatures corresponding to these fixed points, taken as 0 degrees and 100 degrees respectively, we arrive at the empirical centigrade scale of temperature for that thermometer. If the values of the thermometric property P are P_i and P_s at the ice- and steam-points respectively, and if the property has the value P_θ at an unknown temperature θ, the unknown temperature is given by

$$\theta = \frac{100(P_\theta - P_i)}{(P_s - P_i)}$$

on the empirical centigrade scale of this particular thermometer. This equation is illustrated in graphical form in Figure 4.2. It is important to realise that the choice of a different thermometric substance and thermometric property would lead to a different centigrade scale. Agreement between scales occurs only at the two fixed points. This happens because the property may not vary linearly with temperature.

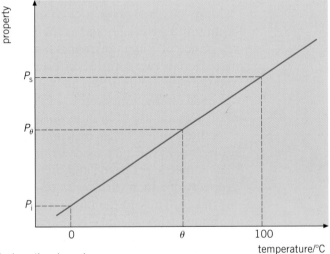

Figure 4.2 Empirical centigrade scale

This situation, with temperature values depending on the type of thermometer on which they are measured, is clearly unsatisfactory for scientific purposes. It is found that the differences between empirical scales are small in the case of thermometers based on gases as thermometric substances. In the constant-volume gas thermometer (Figure 4.3), the pressure of a fixed volume of gas (measured by the height difference h) is used as the thermometric property. The differences between the scales of different gas thermometers become even less as the pressures used are reduced. This is because the lower the pressure of a real gas, the more linear is the variation of pressure or volume. If we set up an empirical centigrade scale for a real gas in a constant-volume thermometer by obtaining the pressures of the gas at the ice-point ($0\,^\circ\text{C}$) and the steam-point ($100\,^\circ\text{C}$) we can extrapolate the graph of pressure p against the centigrade temperature θ to find the temperature at which the pressure of the gas would become zero.

This is shown in Figure 4.4. The extrapolated temperature will be found to be close to −273 degrees on the empirical centigrade constant-volume gas thermometer scale. If the experiment is repeated with lower and lower pressures of gas in the thermometer, the extrapolated temperature tends to a value of −273.15 degrees. This temperature is known as **absolute zero** on the thermodynamic scale of temperature. It does not depend on the properties of any particular substance. At this temperature, any substance has minimum energy (internal energy).

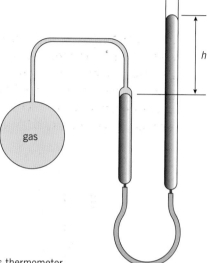

Figure 4.3 Constant-volume gas thermometer

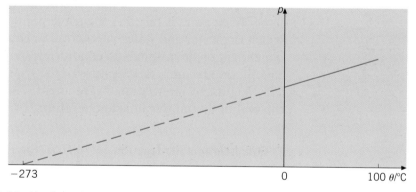

Figure 4.4 Graph of p against θ for constant-volume gas thermometer

Example

The length of the mercury column in a mercury-in-glass thermometer is 25 mm at the ice-point and 180 mm at the steam-point. What is the temperature when the length of the column is 55 mm?

Using the equation $\theta = 100(P_\theta - P_i)/(P_s - P_i)$, the thermometric property P is the length of the mercury column.

Thus, $\theta = 100(55 - 25)/(180 - 25) =$ **19.4 °C** on the centigrade scale of this mercury-in-glass thermometer.

Now it's your turn

A resistance thermometer has resistance $95.2\,\Omega$ at the ice-point and $138.6\,\Omega$ at the steam-point. What resistance would be obtained at a temperature of $19.4\,°C$ on the centigrade scale of this resistance thermometer?

Thermodynamic temperature

As mentioned above, decreasing the pressure of a real gas makes it behave more and more like an ideal gas. For an ideal gas, the relation between pressure p, volume V and temperature T is

$$\frac{pV}{T} = \text{constant}$$

Here T is the **thermodynamic temperature**. We now use the property of an ideal gas to define the thermodynamic scale of temperature. We have already seen from the idea of extrapolating the p, θ graph for a constant-volume gas thermometer that there seems to be a natural zero of temperature, absolute zero. This is used as one of the fixed points of the thermodynamic scale. The upper fixed point is taken as the triple point of water – the temperature at which ice, water and water vapour are in equilibrium. This is found to be less dependent on environmental conditions, such as pressure, than the ice-point. The thermodynamic temperature of the triple point of water is taken as 273.16 units, by international agreement. This defines the kelvin (symbol K), the unit of thermodynamic temperature.

One kelvin is the fraction 1/273.16 of the thermodynamic temperature of the triple point of water.

Thus, if a constant-volume gas thermometer gives a pressure reading of p_{tr} at the triple point, and a pressure reading of p at an unknown temperature T, the unknown temperature (in K) is given by

$$T = 273.16(p/p_{tr})$$

The Celsius scale

Why choose 273.16 as the number of units between the two fixed points of this scale? The reason is that this number will give 100 K between the ice- and steam-points, allowing agreement between the thermodynamic temperature scale and a centigrade scale based on the pressure of an ideal gas. The ideal-gas centigrade scale is based on experiments with real gases at decreasing pressures. This agreement is based on a slightly awkward linking up of the theoretical thermodynamic scale and the empirical constant-volume gas thermometer scale. To avoid this complication, a new scale, the Celsius scale, was defined by international agreement.

> The unit of temperature on the Celsius scale is the degree Celsius (°C), which is exactly equal to the kelvin.

The equation linking temperature θ on the Celsius scale and thermodynamic temperature T is

$$\theta/°C = T/K - 273.15 \quad \text{or} \quad T/K = \theta/°C + 273.15$$

In this equation, θ is measured in °C and T in K. Note that the degree sign (°) always appears with the Celsius symbol, but it is never used with the kelvin symbol K.

Be careful to avoid confusion between the numbers 273.15 and 273.16. Absolute zero on the ideal gas constant-volume scale is −273.15 degrees. The fact that 273.16 occurs in the definition of the kelvin means that the temperature of the triple point of water is 0.01°C. For many purposes, in particular in calculations involving conversions of temperatures from the Celsius to the thermodynamic scale or the other way, it is sufficient to work with the number 273.

Example

The pressure reading of a constant-volume gas thermometer is 1.50×10^4 Pa at the triple point of water. When the bulb is placed in a certain liquid, the pressure is 4.28×10^3 Pa. Find the temperature of the liquid.

We use the equation $T = 273.16(p/p_{tr})$. Because the pressure readings are given to three significant figures only, we can approximate the 273.16 to 273. Thus, $T = 273 \times (4.28 \times 10^3/1.50 \times 10^4) = \textbf{77.9 K}$. (The liquid is liquid nitrogen at its boiling-point.)

Now it's your turn

The pressure reading of a constant-volume gas thermometer is 1.367×10^3 Pa at the steam-point. What is the pressure reading at the triple point of water?

Section 4.1 Summary

- Two regions that are at the same temperature are said to be in thermal equilibrium.
- Empirical centigrade scale of temperature: $\theta = 100(P_\theta - P_i)/(P_s - P_i)$, where θ is the centigrade temperature on the scale of a thermometer based on a property P. P_s, P_i and P_θ are the values of the property at the steam-point, ice-point and at temperature θ respectively.
- The kelvin (K), unit of thermodynamic temperature, is the fraction 1/273.16 of the thermodynamic temperature of the triple point of water.
- Thermodynamic scale of temperature: for a constant-volume gas thermometer, $T = 273.16(p/p_{tr})$, where T is the thermodynamic temperature, p is the pressure reading at temperature T and p_{tr} is the reading at the triple point of water.
- Celsius scale: $\theta = T - 273.15$, where θ is the Celsius temperature (in °C) and T is the thermodynamic temperature (in K).

Section 4.1 Questions

1 The temperature of some water is found to be 23 °C. The water is heated by 15 °C. What are the initial temperature and the temperature rise, measured in kelvin?

2 Three spheres A, B and C are situated close to one another. There is no net transfer of thermal energy between sphere A and sphere C. Likewise, there is no net thermal energy transfer between spheres A and B.

What can be said about the temperatures of spheres B and C? Explain your answer.

3 The temperature of an object is measured using two different types of thermometer.
(a) Explain why the temperatures may be different, when measured on the empirical scales of the two thermometers.
(b) Explain what is meant by the thermodynamic scale of temperature.

4.2 The gas laws

Figure 4.5 Robert Boyle

At the end of Section 4.2 you should be able to:
- state Boyle's law
- select and apply $\dfrac{pV}{T} = \text{constant}$
- state that one mole of any substance contains 6.02×10^{23} particles and that $6.02 \times 10^{23}\,\text{mol}^{-1}$ is the Avogadro constant N_A
- select and solve problems using the ideal gas equation expressed as

  $pV = NkT$ and $pV = nRT$

  where N is the number of atoms and n is the number of moles.

Experiments in the seventeenth and eighteenth centuries showed that the volume, pressure and temperature of a given sample of gas are all related. The precise relation between these quantities and the mass of gas in the sample is called the **equation of state** of the gas. For a given mass of gas, it was found that

> The volume V of the gas is inversely proportional to its pressure p, provided that the temperature is held constant.

Expressed mathematically, this is

$$V \propto 1/p$$

or

$$pV = \text{constant}$$

Another way of writing this equation is

$$p_1 V_1 = p_2 V_2$$

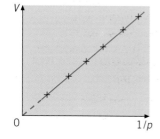

Figure 4.6 Graph of V against $1/p$

where p_1 and V_1 are the initial pressure and volume of the gas, and p_2 and V_2 are the final values after a change of pressure and volume carried out at constant temperature.

A graph of V against $1/p$ is a straight line through the origin (Figure 4.6). This relation is known as **Boyle's law**, after Robert Boyle (1627–91), who first stated this relation as a result of his own experiments.

The effect of temperature on the volume of a gas was investigated by the French scientist Jacques Charles (1746–1823). Charles found that the graph of volume V against temperature θ is a straight line (Figure 4.7). Because gases liquefy when the temperature is reduced, experimental points could not be obtained below the liquefaction temperature. But if the graph was projected backwards, it was found that it cut the temperature axis at about $-273\,°C$.

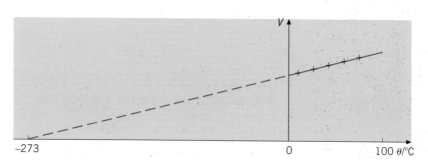

Figure 4.7 Graph of V against θ

In the description of the constant-volume gas thermometer, we have already met the effect of temperature on the pressure of a gas. This effect was investigated by another Frenchman, Joseph Gay-Lussac (1778–1850).

Remember that the graph of pressure p against temperature θ is a straight line, which, if projected like the volume–temperature graph, also meets the temperature axis at about $-273\,°C$ (Figure 4.4). We used this fact to introduce the thermodynamic scale of temperature and the idea of the absolute zero of temperature.

If Celsius temperatures are converted to thermodynamic temperatures, Charles' results are expressed as

> For a given mass of gas maintained at constant pressure, the volume V of the gas is directly proportional to its thermodynamic temperature T.

Expressed mathematically, this is

$$V \propto T$$

This is **Charles' law**. Another way of writing the equation is

$$\frac{V_1}{T_1} = \frac{V_2}{T_2}$$

where V_1 and T_1 are the initial volume and temperature and V_2 and T_2 are the final values. The corresponding relation between pressure and temperature is

> For a given mass of gas maintained at constant volume, the pressure p of the gas is directly proportional to its thermodynamic temperature T.

Expressed mathematically, this is

$$p \propto T$$

This is **Gay-Lussac's law**, or the **law of pressures**. The equation can also be written as

$$\frac{p_1}{T_1} = \frac{p_2}{T_2}$$

where the subscripts 1 and 2 again refer to initial and final values.

We can combine the three gas laws into a single relation between pressure p, volume V and temperature T for a fixed mass of gas. This relation is

$$pV \propto T$$

or

$$\frac{p_1 V_1}{T_1} = \frac{p_2 V_2}{T_2}$$

All the above laws relate to a fixed mass of gas. Another series of experiments could be carried out to find out how the volume of a gas, held at constant pressure and temperature, depends on the mass of gas present. It would be found that the volume is proportional to the mass. This would give the combined relation

$$pV \propto mT$$

where m is the mass of gas, or

$$pV = AmT$$

where A is a constant of proportionality. However, this is not a very useful way of expressing the relation, as the constant A has different values for different gases. We need to find a way of including the quantity of gas. The way to do this is to express the fixed mass of gas in Boyle's, Charles' and Gay-Lussac's laws in terms of the number of moles of gas present.

We have met the mole in Chapter 1 of the AS book, as one of the base SI units.

The **mole** (abbreviated mol) is the amount of substance which contains as many elementary entities as there are atoms in 0.012 kg of carbon-12.

The **Avogadro constant N_A** is the number of atoms in 0.012 kg of carbon-12. That is, it is the number of atoms in a mole of atoms. It has the value 6.02×10^{23} mol^{-1}.

The **relative atomic mass A_r** is the ratio of the mass of an atom to one-twelfth of the mass of an atom of carbon-12. Since the definitions of the mole and the Avogadro constant relate to 0.012 kg (or 12 g) of carbon-12, the relative atomic mass is numerically equal to the mass in grams of a mole of atoms.

Similarly, the **relative molecular mass M_r** is the ratio of the mass of a molecule to one-twelfth of the mass of an atom of carbon-12, and is numerically equal to the mass in grams of a mole of molecules.

The **atomic mass unit u** is one-twelfth of the mass of an atom of carbon-12, and has the value 1.66×10^{-27} kg. (This quantity is also called the unified atomic mass constant.)

Thus, a simpler way of finding the mass of one mole of a particular element is to take its nucleon number (see Chapter 7) expressed in grams. For example, the nucleon number of argon (^{40}Ar) is 40. One mole of argon then has mass 40 g.

Returning to the combination of the three gas laws and using the number n of moles of the gas, we have

$$pV \propto nT$$

or, putting in a new constant of proportionality R,

$$pV = nRT$$

R is called the **molar gas constant** (sometimes the **universal gas constant**, because it has the same value for all gases). It has the value $8.3\,\mathrm{J\,K^{-1}mol^{-1}}$. Sometimes the equation $pV = nRT$ is expressed in the form

$$pV_{\mathrm{m}} = RT$$

where the molar volume V_{m} is the volume occupied by one mole of the gas. Another version is

$$pV = NkT$$

where N is the number of molecules in the gas and k is a constant called the **Boltzmann constant**. It has the value $1.38 \times 10^{-23}\,\mathrm{J\,K^{-1}}$. The molar gas constant R and the Boltzmann constant k are connected through the Avogadro constant N_{A}

$$R = kN_{\mathrm{A}}$$

Strictly speaking, the laws of Boyle, Charles and Gay-Lussac are not really laws, as their validity is restricted. They are accurate for real gases only if the pressure of the sample is not too great, and if the gas is well above its liquefaction temperature. But they can be used to define an 'ideal' gas.

An **ideal gas** is one which obeys the equation of state

$$pV = nRT$$

at all pressures, volumes and temperatures.

We have already used the ideal gas equation to introduce the idea of thermodynamic temperature. For approximate calculations, the ideal gas equation can be used with real gases.

Examples

1 Find the volume occupied by 1 mole of air at standard temperature and pressure (273 K and 1.01×10^5 Pa), taking R as $8.3\,\text{J K}^{-1}\,\text{mol}^{-1}$ for air.

Since $pV = nRT$, $V = nRT/p$. Substituting the values $n = 1$ mol, $R = 8.3\,\text{J K}^{-1}\,\text{mol}^{-1}$, $T = 273$ K and $p = 1.01 \times 10^5$ Pa, $V = (1 \times 8.3 \times 273)/1.01 \times 10^5 = \mathbf{2.24 \times 10^{-2}\,m^3}$.

2 Find the number of molecules per cubic metre of air at standard temperature and pressure.

We have just shown that the volume occupied by one mole of air at standard temperature and pressure is $2.24 \times 10^{-2}\,\text{m}^3$. One mole of air contains N_A molecules, where N_A is the Avogadro constant $(6.02 \times 10^{23}\,\text{mol}^{-1})$. Thus the number of molecules per cubic metre of air is $6.02 \times 10^{23}/2.24 \times 10^{-2} = \mathbf{2.69 \times 10^{25}\,m^{-3}}$.

Note: It is useful to remember these two quantities, the *molar volume* of a gas and the *number density* of molecules in it. They give an idea of the relatively small volume occupied by a mole of gas at standard temperature and pressure (a cube of side about 28 cm), and the enormous number of molecules in every cubic metre of a gas under these conditions.

3 A syringe contains $25 \times 10^{-6}\,\text{m}^3$ of helium gas at a temperature of $20\,°C$ and a pressure of 5.0×10^4 Pa. The temperature is increased to $400\,°C$ and the pressure on the syringe is increased to 2.4×10^5 Pa. Find the new volume of gas in the syringe.

Boyle's law, $pV = $ constant, and Charles' law, $V/T = $ constant, may be combined to give $pV/T = $ constant. This is written in the form $p_1V_1/T_1 = p_2V_2/T_2$ and re-arranged as $V_2 = p_1V_1T_2/p_2T_1$. Substituting the values $p_1 = 5.0 \times 10^4$ Pa, $V_1 = 25 \times 10^{-6}\,\text{m}^3$, $T_2 = 673$ K, $p_2 = 2.4 \times 10^5$ Pa, $T_1 = 293$ K (again note that temperatures are converted from °C to K),

$V_2 = 5.0 \times 10^4 \times 25 \times 10^{-6} \times 673/2.4 \times 10^5 \times 293 = \mathbf{12 \times 10^{-6}\,m^3}$.

Now it's your turn

1 The number of molecules per cubic metre of air at standard temperature and pressure is about $2.7 \times 10^{25}\,\text{m}^{-3}$. What is the average separation of these molecules?

2 The mean mass of one mole of air (which is made up mainly of nitrogen, oxygen and argon) is 0.029 kg. What is the density of air at standard temperature and pressure (273 K and 1.01×10^5 Pa)?

3 The volume of a sample of gas is $3.2 \times 10^{-2}\,\text{m}^3$ when the pressure is 8.6×10^4 Pa and the temperature is $27\,°C$. How many moles of gas are there in the sample? How many molecules? What is the number of molecules per cubic metre?

4 A sample of air has volume $15 \times 10^{-6} \, \text{m}^3$ when the pressure is $5.0 \times 10^5 \, \text{Pa}$. The pressure is reduced to $3.0 \times 10^5 \, \text{Pa}$, without changing the temperature. What is the new volume?

5 A sample of gas, originally at standard temperature and pressure (273 K and $1.01 \times 10^5 \, \text{Pa}$), has volume $3.0 \times 10^{-5} \, \text{m}^3$ under these conditions. When the pressure is increased to $4.0 \times 10^5 \, \text{Pa}$, the temperature rises to $40 \, °\text{C}$. What is the new volume?

Section 4.2 Summary

- Boyle's law states that for a fixed mass of gas at constant temperature, the product of the pressure of a gas and its volume is a constant: pV = constant or $p_1V_1 = p_2V_2$.
- Charles' law states that if the pressure of a fixed mass of gas is kept constant, the ratio of volume to thermodynamic temperature is a constant: V/T = constant or $V_1/T_1 = V_2/T_2$.
- Gay-Lussac's law (the law of pressures) states that if the volume of a fixed mass of gas is kept constant, the ratio of pressure to thermodynamic temperature is a constant: p/T = constant or $p_1/T_1 = p_2/T_2$.
- The equation of state for an ideal gas relates the pressure p, volume V and thermodynamic temperature T of n moles of gas: $pV = nRT$ where $R = 8.3 \, \text{J K}^{-1} \, \text{mol}^{-1}$, the molar gas constant.
- For N molecules of gas: $pV = NkT$ where $k = 1.38 \times 10^{-23} \, \text{J K}^{-1}$, the Boltzmann constant.
- The relation between R and k is: $R = kN_A$ where $N_A = 6.02 \times 10^{23} \, \text{mol}^{-1}$, the Avogadro constant (the number of atoms in 0.012 kg of carbon-12).
- Another expression of the ideal gas equation (or of Boyle's law, Charles' law and Gay-Lussac's law combined) is: pV/T = constant or $p_1V_1/T_1 = p_2V_2/T_2$.

Section 4.2 Questions

1 On a day when the atmospheric pressure is 102 kPa and the temperature is $8 \, °\text{C}$, the pressure in a car tyre is 190 kPa above atmospheric pressure. After a long journey the temperature of the air in the tyre rises to $29 \, °\text{C}$. Calculate the pressure above atmospheric of the air in the tyre at $29 \, °\text{C}$. Assume that the volume of the tyre remains constant.

2 A helium-filled balloon is released at ground level, where the temperature is $17 \, °\text{C}$ and the pressure is 1.0 atmosphere. The balloon rises to a height of 2.5 km, where the pressure is 0.75 atmospheres and the temperature is $5 \, °\text{C}$. Calculate the ratio of the volume of the balloon at 2.5 km to that at ground level.

3 A metal box in the form of a cube of side 30 cm is filled with air at atmospheric pressure ($1.01 \times 10^5 \, \text{Pa}$) at $15 \, °\text{C}$. The box is sealed and heated in an oven to $200 \, °\text{C}$. Calculate the net force on each side of the box.

4.3 Solid, liquid and gas

At the end of Section 4.3 you should be able to:

■ describe solids, liquids and gases in terms of the spacing, ordering and motion of atoms or molecules
■ describe a simple kinetic model for solids, liquids and gases
■ describe using a simple kinetic model for matter the terms *melting*, *boiling* and *evaporation*
■ define the term *pressure* and use the kinetic model to explain the pressure exerted by gases.

We can think of all matter as existing as a solid, a liquid, a gas or a plasma. A plasma is formed at very high temperatures and will not be discussed in this book.

We have seen that all matter consists of atoms or groups of atoms called molecules. In the **solid state**, these atoms or molecules are held in position by strong rigid forces. This explains why a solid is characterised as having a fixed volume and a fixed shape. The atoms or molecules vibrate about their fixed positions in the solid. As the temperature of the solid increases, so the energy of vibration increases. The increased energy of vibration results in

■ the amplitude of vibration of the atoms increasing
■ the separation of the atoms increasing
■ the forces between the atoms (the bonds) decreasing.

These effects are observed as an expansion of the solid on heating and the solid becoming more pliable.

Example

Assuming that the atoms of iron are arranged in a simple cubic lattice as shown in Figure 4.8, estimate the spacing between atoms of iron.

The density of iron is $7900 \, \text{kg m}^{-3}$ and its molar mass is $56 \times 10^{-3} \, \text{kg}$.

One mole contains 6.0×10^{23} atoms.

$1.0 \, \text{m}^3$ has mass $7900 \, \text{kg}$ which is $7900/(56 \times 10^{-3}) = 1.4 \times 10^5 \, \text{mol}$.

Therefore,

$1.0 \, \text{m}^3$ contains $1.4 \times 10^5 \times 6.0 \times 10^{23}$
$= 8.5 \times 10^{28}$ atoms

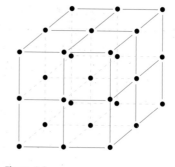

Figure 4.8

Volume 'occupied' by one atom
$= 1/(8.5 \times 10^{28}) = 1.2 \times 10^{-29} \, \text{m}^3$

Separation of atoms $= \sqrt[3]{1.2 \times 10^{-29}} = 2 \times 10^{-10} \, \text{m}$.

As the temperature rises, there comes a point where the vibrations of the atoms are large enough to enable the atoms or molecules to break free from the rigid forces between them. In order to break the bonds, energy must be supplied. However, this energy does not increase the vibrational energy and, therefore, does not change the temperature of the solid. The atoms or molecules break free of the rigid forces and the solid is said to **melt**. In melting, the solid changes from solid to **liquid** as energy is supplied that does not change the temperature – **latent heat of fusion**.

A liquid is characterised as having a fixed volume but no fixed shape. The atoms or molecules in the liquid state are still held close together but the forces are no longer rigid. The atoms or molecules are free to move about within the body of the liquid. Evidence for this is provided by the fact that solids and liquids have approximately the same density, meaning that the atoms and molecules are held at approximately the same separation in both states. However, because the forces are not rigid, the atoms or molecules can move within the body of the liquid, thus meaning that the liquid can flow.

As the temperature of the liquid rises, the energy of the atoms or molecules increases and thus they have greater kinetic energy. On average, the atoms or molecules move faster. The atoms or molecules move randomly and some will be moving at a speed greater than average towards the surface of the liquid. These atoms or molecules that are moving at speeds greater than the average can escape through the surface, overcoming the attractive forces between them. The atoms or molecules enter the space above the liquid, forming a **gas** or **vapour**. The process is known as **evaporation**.

Example

When water evaporates at normal atmospheric pressure, $1.0\,\mathrm{cm^3}$ of liquid becomes $1600\,\mathrm{cm^3}$ of vapour.

Calculate the ratio $\dfrac{\text{average separation of molecules in vapour state}}{\text{average separation of molecules in liquid state}}$.

Let the average separation in the liquid state be d.

Then, if there are N molecules in $1.0\,\mathrm{cm^3}$ of liquid

$Nd^3 = 1$.

If the average spacing in the vapour state is D, then, since there will be the same number of molecules in the vapour as in the liquid,

$ND^3 = 1600$

So, $D^3/d^3 = 1600$

$D/d = \sqrt[3]{1600} \approx 12$

During evaporation, only the more energetic atoms or molecules escape the surface. Thus, the average kinetic energy of those remaining in the liquid state decreases. Since average kinetic energy and temperature are related, it can be concluded that evaporation produces cooling. To maintain a constant temperature, heat must be supplied to the liquid. Since this heating does not change the temperature of the liquid, the heat is known as **latent heat of vaporisation**.

Evaporation occurs from the surface of the liquid and occurs at all temperatures.

As the temperature of the liquid is raised, there comes a point where the pressure of the vapour above the liquid is equal to the external pressure. At this temperature the liquid **boils**. Boiling is a change of state from liquid to gas (or vapour) that occurs in the body of the liquid and at one temperature – the boiling point.

The atoms or molecules in the gas or vapour state have negligible forces of attraction between them. Thus, they can move about randomly. Their average separation is about ten times greater than in the liquid or solid state. As a result, gases are much less dense than liquids or solids and they do not have a fixed volume or shape.

Since the atoms or molecules in a gas move randomly, they collide with each other and with the walls of the containing vessel. These collisions are elastic. When an atom or molecule collides with the wall of the containing vessel, there is a change in momentum of the gas atom and the wall experiences a small force for a short time (an impulse). There are many collisions per unit time and thus the wall experiences a constant force.

Pressure is defined as force per unit area.

$$\text{pressure} = \frac{\text{force}}{\text{area}}$$

Pressure is measured in newtons per square metre ($N\,m^{-2}$) or pascal (Pa).

Since the gas atoms or molecules exert a force on the wall of the containing vessel, then the force per unit area on the wall is known as the gas pressure.

The model of solids, liquids and gases outlined above that is based on the behaviour of atoms or molecules is known as a **simple kinetic model for matter**. Throughout the explanation of the theory, reference was made to properties of solids, liquids and gases (e.g. density, constancy of shape) to support the theory. The question arises as to whether there is any evidence to support the belief that, in a gas, the atoms or molecules move randomly.

Section 4.3 Summary

- Solids have a fixed volume and fixed shape. There are strong rigid interatomic forces.
- Liquids have a fixed volume but no fixed shape. The spacing of atoms is about the same as in the solid state but the interatomic forces do not hold the atoms in fixed positions.
- Gases have no fixed volume or fixed shape. Distances between atoms are about ten times greater than in liquids or solids. Interatomic forces are very small and are zero in the case of an ideal gas.
- Melting, boiling and evaporation are all examples of changes of state or phase (solid to liquid, and liquid to vapour or gas).
- Melting, evaporation and boiling require an input of energy (latent heat) to overcome interatomic forces.
- Evaporation takes place at the surface of a liquid, boiling takes place in the body of the liquid.
- Boiling takes place at a fixed temperature for a particular external (atmospheric) pressure but evaporation occurs at all temperatures.

Section 4.3 Questions

1 By reference to the forces between atoms or molecules, explain why, in general, solids have a fixed volume and shape, whereas liquids have a fixed volume but no fixed shape.

2 Use the ideas about the motion of molecules to explain why evaporation of a liquid produces cooling.

4.4 A microscopic model of a gas

At the end of Section 4.4 you should be able to:

- describe an experiment that demonstrates Brownian motion and discuss the evidence for the movement of molecules provided by such an experiment
- explain that the mean translational kinetic energy of an atom of an ideal gas is directly proportional to the temperature of the gas in kelvin
- select and apply the equation $\langle E_k \rangle = \frac{3}{2}kT$ for the mean translational kinetic energy of atoms.

In 1827 the biologist Robert Brown was observing, under a microscope, tiny pollen grains suspended in water. He saw that the grains were always in a jerky, haphazard motion, even though the water was completely still. This movement is now called **Brownian motion**. If we make the assumption that the water molecules are in rapid, random motion, then it is easy to see how the pollen grains could move so jerkily under the random bombardment,

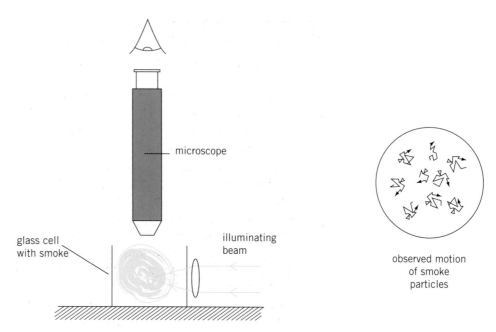

Figure 4.9 Observing Brownian motion

from all sides, of the water molecules. We can also reproduce Brownian motion by observing the motion of tiny soot particles in smoke. These particles, too, move in the jerky manner of Brown's pollen grains. Thus it seems that the molecules of both gases and liquids are in the rapid, random motion as required by the molecular model. Nearly a century later, Einstein made a theoretical analysis of Brownian motion and, from the masses and distances moved by the suspended particles, was able to estimate that the diameter of a typical atom is of the order of 10^{-10} m.

The kinetic theory

An explanation of how a gas exerts a pressure was developed by Robert Boyle in the seventeenth century, and in greater detail by Daniel Bernoulli in the eighteenth century. As already oulined, the basic idea was that the gas consists of atoms or molecules moving about at great speed (later visualised by Robert Brown). The gas exerts a pressure on the walls of its container because of the continued impacts of the molecules with the walls. In the kinetic theory of an ideal gas the following assumptions are made:

1 All molecules behave as identical, hard, perfectly elastic spheres.
2 The volume of the molecules is negligible compared with the volume of the containing vessel.
3 There are no forces of attraction or repulsion between molecules.
4 There are many molecules, all moving randomly.

Using these assumptions and also the ideal gas equation, it can be shown that the mean (average) translational kinetic energy $\langle E_k \rangle$ of a molecule in an ideal gas is proportional to the absolute (kelvin) temperature. That is

$$\langle E_k \rangle \propto T$$

The constant in the expression above is $\frac{3}{2}k$. That is

$$\langle E_k \rangle = \frac{3}{2}kT$$

where k is the Boltzmann constant ($1.38 \times 10^{-23}\,\text{J K}^{-1}$).

This is an important result. We have obtained a relation between the mean kinetic energy of a molecule in a gas and the temperature of the gas.

For a given gas, the higher the temperature, the faster the molecules move.

Examples

Find the total kinetic energy of the molecules in one mole of an ideal gas at standard temperature (273 K).

We know that the mean kinetic energy of one molecule is $\frac{3}{2}kT$.

For one mole of molecules, that is N_A molecules, the energy is

$$\frac{3}{2}N_A kT \text{ or } \frac{3}{2}RT$$

Substituting the values $R = 8.3\,\text{J K}^{-1}\,\text{mol}^{-1}$ and $T = 273\,\text{K}$, we have

$$\langle E_k \rangle = \frac{3}{2} \times 8.3 \times 273 = \textbf{3400 J mol}^{-1}.$$

Now it's your turn

Find the mean kinetic energy of a molecule in an ideal gas at a temperature of 500 K.

Section 4.4 Summary

■ Assumptions of the kinetic theory of gases:

    1   Molecules behave as identical, hard, perfectly elastic spheres.
    2   Volume of the molecules is negligible compared with the volume of the containing vessel.
    3   There are no forces of attraction or repulsion between molecules.
    4   There are many molecules, all moving randomly.

■ Mean (average) kinetic energy of a molecule: $\langle E_k \rangle = \frac{3}{2}kT$

Section 4.4 Questions

1 Explain, on the basis of the kinetic theory of gases:
  (a) why a gas exerts a pressure on the walls of its containing vessel,
  (b) the effect on the pressure of heating the gas at constant volume,
  (c) the effect on the pressure of reducing the volume of the gas at constant temperature.

2 In a real gas, there are forces of attraction between the molecules of the gas. Suggest and explain the effect of such forces on the pressure exerted as compared to the predicted pressure, assuming the gas to be ideal.

4.5 # 4.5 Internal energy

At the end of Section 4.5 you should be able to:

- define *internal energy* as the sum of the random distribution of kinetic and potential energies associated with the molecules of a system
- explain that the rise in temperature of a body leads to an increase in its internal energy
- explain that a change of state for a substance leads to changes in its internal energy but not its temperature.

We have seen that the molecules of an ideal gas possess kinetic energy, and that this kinetic energy is proportional to the thermodynamic temperature of the gas. The sum of the kinetic energies of all the molecules, due to their random motion, is called the **internal energy** of the ideal gas. Not all molecules have the same kinetic energy, because they are moving with different speeds, but the sum of all the kinetic energies will be a constant if the gas is kept at a constant temperature.

For a real gas, the situation is a little more complicated. Because the molecules of a real gas exert forces on each other, at any instant there will be a certain potential energy associated with the positions the molecules occupy in space. Because the molecules are moving randomly, the potential energy of a given molecule will also vary randomly. But at a given temperature the total potential energy of all the molecules will remain constant. If the temperature changes, the total potential energy will also change. Furthermore, the molecules of a real gas collide with each other, and will interchange kinetic energy during the collisions.

> The internal energy is given by the sum of the random distribution of the potential energies and the kinetic energies of all the molecules.

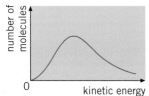

Figure 4.10 Distribution of molecular kinetic energies

It is important to realise that looking at a single molecule will give us very little information. Its kinetic energy will be changing all the time as it collides with other gas molecules, and its potential energy is also changing as its position relative to the other molecules in the gas changes. This single molecule has a kinetic energy which is part of the very wide range of kinetic energies of the molecules of the gas. We say that there is a *distribution* of molecular kinetic energies. The distribution is illustrated in Figure 4.10. Similarly, there is a distribution of molecular potential energies. But by adding up the kinetic and potential energies of all the molecules in the gas, the random nature of the kinetic and potential energies of the single molecule is removed.

The idea of internal energy can be extended to all states of matter. In a liquid, intermolecular forces are stronger as the molecules are closer together, so the potential energy contribution to internal energy becomes more significant. The kinetic energy contribution is still due to the random motion of the molecules in the liquid. We can think of a solid as being made up of atoms or molecules which oscillate (vibrate) about equilibrium positions. Here, the potential energy contribution is caused by the strong binding forces between atoms, and the kinetic energy contribution is due to the motion of the vibrating atoms.

The concept of internal energy is particularly useful as it helps us to distinguish between temperature and heat. Using an ideal gas as an example, temperature is a measure of the *average* kinetic energy of the molecules. It therefore does not depend on how many molecules are present in the gas. Internal energy (again for an ideal gas), however, is the *total* kinetic energy of the molecules, and clearly does depend on how many molecules there are. In general heat refers to a *transfer* of energy from one substance to another, often as a result of a temperature difference. If two objects at different temperatures are placed in contact, there will be a flow of heat from the object at the higher temperature to the one at the lower. The direction of heat flow is determined by the difference in temperature, not by any difference in internal energies. If 10 g of a liquid at 30 °C is placed in contact, or mixed into, 100 g of the same liquid at 20 °C, the direction of heat flow is from the liquid at 30 °C to the liquid at 20 °C, even though the liquid at 20 °C has a greater internal energy than the smaller mass of liquid at 30 °C.

We have seen that when a substance changes from solid to liquid (melting or fusion) or from liquid to vapour (evaporation or boiling), intermolecular bonds are broken. This increases the potential energy component of the internal energy. During the change of state, the temperature does not change. Therefore the kinetic energy of the molecules does not change. This means that, although the temperature does not change, a change of state involves a change in internal energy.

We know that the internal energy of a system is the total energy, kinetic and potential, of the various parts of the system. For a system consisting of an ideal gas, the internal energy is simply the total kinetic energy of all the atoms or molecules of the gas. For such a system, we would expect the internal energy to increase if heat were added to the gas, or if work were done on it. In both cases we are adding energy to the system. By the law of conservation of energy, this energy cannot just disappear; it must be transformed to another type of energy. It appears as an increase in the internal energy of the gas; that is, the total kinetic energy of the molecules is increased. But if the *total* kinetic energy of the molecules increases, their *average* kinetic energy is also increased. Because the average kinetic energy is a measure of temperature, the addition of energy to the ideal gas shows up as an increase in its temperature.

Section 4.5 Summary

- The internal energy of a substance is the sum of the random kinetic and potential energies of its atoms or molecules.
- For an ideal gas, the total internal energy is the total kinetic energy of its molecules.
- A rise in temperature of a substance leads to an increase in internal energy.
- A change of state of a substance leads to a change in internal energy without any change of temperature.

Section 4.5 Questions

1 Explain why, in a change where no thermal energy is allowed to enter or leave a system, the work done on the system is equal to the increase in the internal energy of the system.

2 Some water is released and observed in a waterfall as it descends into a pool at the foot of the waterfall.
   (a) Just before the water hits the still water in the pool, it has kinetic energy. Suggest and explain whether this kinetic energy is 'ordered' or 'random' kinetic energy of the molecules of water.
   (b) The water in the pool at the bottom of the waterfall is stationary. State the conversion that has occurred in the kinetic energy of the water and hence explain why the temperature of the water at the top of the waterfall is different to that in the pool.

4.6 Thermal properties of materials

At the end of Section 4.6 you should be able to:
- define and apply the concept of specific heat capacity
- select and apply the equation $E = mc\Delta\theta$
- describe an electrical experiment to determine the specific heat capacity of a solid or a liquid
- describe what is meant by the terms *latent heat of fusion* and *latent heat of vaporisation*.

In this section, we will be looking at the effect of heating on temperature.

Specific heat capacity

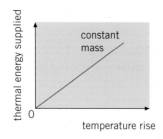

Figure 4.11

When a solid, a liquid or a gas is heated, its temperature rises. Plotting a graph of thermal (heat) energy supplied against temperature rise (see Figure 4.11), it is seen that the temperature rise $\Delta\theta$ is proportional to the heat energy ΔQ supplied, for a particular mass of a particular substance.

$$\Delta Q \propto \Delta\theta$$

Similarly, the heat energy required to produce a particular temperature rise is proportional to the mass m of the substance being heated (see Figure 4.12).

$$\Delta Q \propto m$$

Combining these two relations gives

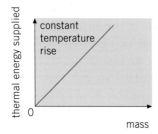

Figure 4.12

$$\Delta Q \propto m\Delta\theta$$

$$\Delta Q = mc\Delta\theta$$

where c is the constant of proportionality known as the **specific heat capacity** of the substance. In this case, *specific* means *per unit mass*.

> The numerical value of the specific heat capacity of a substance is the quantity of heat energy required to raise the temperature of unit mass of the substance by one degree.

The SI unit of specific heat capacity is $J\,kg^{-1}\,K^{-1}$. The unit of specific heat capacity is *not* the joule and this is why, in the definition of specific heat capacity, it is important to make reference to the *numerical value*. Specific heat capacity is different for different substances. Some values are given in Table 4.1.

Table 4.1 Values of specific heat capacity for different materials

material	specific heat capacity/$J\,kg^{-1}\,K^{-1}$
ethanol	2500
glycerol	2420
ice	2100
mercury	140
water	4200
aluminium	913
copper	390
glass	640

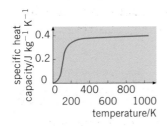

Figure 4.13

It should be noted that, for relatively small changes in temperature, specific heat capacity is approximately constant. However, over a wide range of temperature, the value for a substance may vary considerably (see Figure 4.13). Unless stated otherwise, specific heat capacity is assumed to be constant.

Determination of specific heat capacity

The metal whose specific heat capacity is to be found is in the form of a cylinder. The mass of the cylinder is m. The cylinder has two holes drilled in it so that an immersion heater and a thermometer may be inserted into the metal cylinder. The apparatus is set up as shown in Figure 4.14.

The initial temperature θ_I of the metal is noted. The heater is switched on and a clock started.

The variable resistor is adjusted to keep the current constant. The average values of the voltage V and the current I are noted.

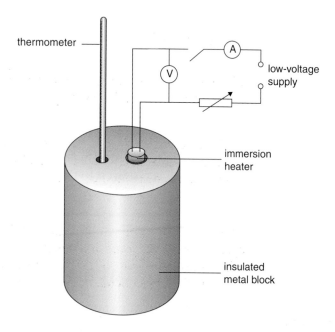

thermometer

A

V

low-voltage
supply

immersion
heater

insulated
metal block

Figure 4.14

When the temperature has risen by about 20 °C, the current is switched off and the time t for which the block was heated is noted. The highest temperature θ_H recorded by the thermometer is noted.

Assuming there are no heat losses to the surroundings,

heat supplied = heat gained by block

$$VIt = mc(\theta_H - \theta_I)$$

Hence, the specific heat capacity c can be determined.

To determine the specific heat capacity of a liquid, a known mass of the liquid can be contained in a polystyrene cup. The same method is used as for a solid but the liquid should be stirred. The polystyrene cup absorbs negligible heat and acts as insulation.

Example

Calculate the quantity of heat energy required to raise the temperature of a mass of 810 g of aluminium from 20 °C to 75 °C. The specific heat capacity of aluminium is $910 \, \text{J} \, \text{kg}^{-1} \text{K}^{-1}$.

$$
\begin{aligned}
\text{heat energy required} &= m \times c \times \Delta\theta \\
&= \frac{810}{1000} \times 910 \times (75 - 20) \\
&= \mathbf{4.1 \times 10^4 \ J}
\end{aligned}
$$

Now it's your turn

1 Calculate the heat energy gained or lost for the following temperature changes. Use Table 4.1 to obtain values for specific heat capacity.
   (a) 45 g of copper heated from 10 °C to 90 °C
   (b) 1.3 kg of ice at 0 °C cooled to −15 °C

2 Calculate the specific heat capacity of water given that 0.20 MJ of energy is required to raise the temperature of a mass of 600 g of water by 80 K.

Specific latent heat

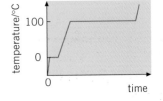

Figure 4.15

Figure 4.15 illustrates how the temperature of a mass of water varies with time when it is heated at a constant rate. At times when the substance is changing phase (ice to water or water to steam), heat energy is being supplied without any change of temperature. Because the heat energy does not change the temperature of the substance, it is said to be **latent** (i.e. hidden). The latent heat required to melt (fuse) a solid is known as **latent heat of fusion**.

> The numerical value of the **specific latent heat of fusion** is the quantity of heat energy required to convert unit mass of solid to liquid without any change in temperature.

The SI unit of specific latent heat of fusion is J kg^{-1}. For a substance with latent heat of fusion L_f, the quantity of heat energy ΔQ required to fuse (melt) a mass m of solid is given by

$$\Delta Q = mL_f$$

The latent heat required to vaporise a liquid is referred to as **latent heat of vaporisation**.

> The numerical value of the **specific latent heat of vaporisation** is the quantity of heat energy required to convert unit mass of liquid to vapour without any change in temperature.

The SI unit of specific latent heat of vaporisation is the same as that for fusion i.e. J kg^{-1}. For a substance with latent heat of vaporisation L_v, the quantity of heat energy ΔQ required to vaporise a mass m of liquid is given by

$$\Delta Q = mL_v$$

When a vapour condenses (vapour becomes liquid), the latent heat of vaporisation is released. Similarly, when a liquid solidifies (liquid becomes solid), the latent heat of fusion is released. Some values of specific latent heat of fusion and of vaporisation are given in Table 4.2.

Table 4.2 Values of specific latent heat

material	specific latent heat of fusion/kJ kg^{-1}	specific latent heat of vaporisation/kJ kg^{-1}
ice/water	330	2260
ethanol	108	840
copper	205	4840
sulfur	38.1	

Example

Use the information given in Table 4.2 to determine the heat energy required to melt 50 g of sulfur at its normal melting point.

$$heat\ energy\ required = m \times L_f$$
$$= \frac{50}{1000} \times 38.1 \times 1000$$
$$= \mathbf{1900\ J}$$

Now it's your turn

Where appropriate, use the information given in Table 4.2.
(a) Calculate the heat energy required to:
  (i) melt 50 g of ice at 0 °C,
  (ii) evaporate 50 g of water at 100 °C.
(b) Using your answers to (a), determine how many times more energy is required to evaporate a mass of water than to melt the same mass of ice.

Exchanges of heat energy

When a hot object and a cold object come into contact, heat energy passes from the hot object to the cold one so that the two objects reach the same temperature. The law of conservation of energy applies in that the heat energy gained by the cold object is equal to the heat energy lost by the hot object. This does, of course, assume that no energy is lost to the surroundings. This simplification enables temperatures to be calculated.

Example

A mass of $0.30\,\text{kg}$ of water at $95\,^\circ\text{C}$ is mixed with $0.50\,\text{kg}$ of water at $20\,^\circ\text{C}$. Calculate the final temperature of the water, given that the specific heat capacity of water is $4200\,\text{J}\,\text{kg}^{-1}\,\text{K}^{-1}$.

Hint: always start by writing out a word equation containing all the gains and losses of heat energy.

heat energy lost by hot water = heat energy gained by cold water
$$(m \times c \times \Delta\theta_1) = (M \times c \times \Delta\theta_2)$$
$$0.30 \times 4200 \times (95 - t) = 0.50 \times 4200 \times (t - 20)$$

where t is the final temperature of the water.

$$1260 \times (95 - t) = 2100 \times (t - 20)$$
$$119\,700 - 1260t = 2100t - 42\,000$$
$$161\,700 = 3360t$$
$$\boldsymbol{t = 48\,^\circ\text{C}}$$

Now it's your turn

Where appropriate use the data in Table 4.1.

A lump of copper of mass $120\,\text{g}$ is heated in a gas flame. It is then transferred to a mass of $450\,\text{g}$ of water, initially at $20\,^\circ\text{C}$. The final temperature of the copper and the water is $31\,^\circ\text{C}$. Calculate the temperature of the gas flame.

Section 4.6 Summary

- Specific heat capacity is numerically equal to the heat energy required to raise the temperature of unit mass of substance by one degree.
- The heat energy ΔQ required to raise the temperature of a mass m of substance of specific heat capacity c by an amount $\Delta\theta$ is given by the expression: $\Delta Q = mc\Delta\theta$.
- Specific heat capacity has the SI unit of $\text{J}\,\text{kg}^{-1}\,\text{K}^{-1}$.
- Specific latent heat of fusion is numerically equal to the quantity of heat energy required to convert unit mass of solid to liquid without any change in temperature.
- Specific latent heat of vaporisation is numerically equal to the quantity of heat energy required to convert unit mass of liquid to vapour without any change in temperature.
- Specific latent heat has the SI unit of $\text{J}\,\text{kg}^{-1}$.

Section 4.6 Questions

1 A polystyrene cup contains a mass of 95 g of water at 27 °C. A small lump of copper of mass 32 g is heated in a Bunsen flame and then transferred to the water. The temperature of the Bunsen flame is 290 °C.
The specific heat capacities of water and copper are $4.2 \, \mathrm{J\,g^{-1}\,K^{-1}}$ and $0.39 \, \mathrm{J\,g^{-1}\,K^{-1}}$ respectively.
(a) Calculate the final temperature of the water in the polystyrene cup.
(b) State two assumptions that you make in your calculation.

2 Use ideas about the molecules in gases and in solids to suggest why the specific heat capacity of a gas is much less than that of a solid.

3 A 'hot water' bottle is used to store thermal energy in heated water. Suggest why a mass of water provides a better store of thermal energy than an equal mass of copper.

Exam-style Questions

1 Consider one mole of gas at standard temperature and pressure (273 K and 1.01×10^5 Pa).
(a) Calculate:
  (i) the number of molecules there are per cubic metre of this gas (this is known as the Loschmidt number N_L),
  (ii) the average distance apart of the molecules.
(b) If the diameter of a molecule of this gas is about 2.5×10^{-10} m, estimate the fraction of each cubic metre occupied by matter.

2 A volume of $1.50 \times 10^3 \, \mathrm{m^3}$ of hydrogen, at a pressure of 2.00×10^5 Pa and a temperature of 30 °C, is heated until both volume and pressure are doubled. Calculate:
(a) the final temperature,
(b) the mass of hydrogen. (mass of one mole of hydrogen = 2.00×10^{-3} kg)

3 Two containers, each of equal, constant volume, are connected by a tube of negligible volume. When the temperature of both containers is 27 °C, the pressure of the gas in the containers is 2.1×10^5 Pa. One container is heated to a temperature of 127 °C, whilst the other is maintained at 27 °C. Calculate the final pressure in the containers.

4 The kinetic theory leads to the expression

$$\langle E_k \rangle = \tfrac{3}{2} kT$$

for the mean (average) kinetic energy of a molecule of a gas. The constant k does not depend on the type of molecule. Can this result really be true both for hydrogen and chlorine? (The mass of a chlorine molecule is about 35 times that of a hydrogen molecule.) Explain why.

5 The mass of an oxygen molecule is 16 times the mass of a hydrogen molecule. Equal masses of hydrogen and oxygen gases are mixed, and are allowed to come to equilibrium. Calculate:
(a) the ratio of numbers of molecules,
(b) the ratio of the average kinetic energy per molecule,
(c) the ratio of pressures exerted on the walls of the container.

5.Fields

In Chapter 3 we learnt about the law of gravitation, the law which governs the force between masses, and about gravitational fields. The aim of this chapter is to investigate the nature of electric forces and fields. We shall find some similarities between electric fields and gravitational fields, and some differences. Electric forces hold electrons in atoms, and bind atoms together in molecules and in solids. Gravitational forces are responsible for holding the Moon in orbit around the Earth. Both gravitational and electric fields are fields in which forces act over a distance. In this study, we shall develop an appreciation of both sorts of field in our physical world. As an example of the application of electric fields in circuit components, we shall study the operation, function and applications of capacitors in detail.

5.1 Electric charges

At the end of Section 5.1 you should be able to:
- recall and use Coulomb's law in the form $F = Q_1 Q_2 / 4\pi\epsilon_0 r^2$ for the force between two point charges in free space or air
- show an understanding of the concept of an electric field as an example of a field of force and define electric field strength as force per unit positive charge
- represent an electric field by means of field lines
- recall and use $E = V/d$ to calculate the field strength of the uniform field between charged parallel plates in terms of potential difference and separation
- calculate the forces on charges in uniform electric fields
- explain the effect of a uniform electric field on the motion of charged particles
- recall and use $E = Q/4\pi\epsilon_0 r^2$ for the field strength of a point charge in free space or air
- note the analogy between electric and gravitational forces and fields, describing the similarities and differences.

Some effects of static electricity are familiar in everyday life. For example, a balloon rubbed on a jumper will stick to a wall, a TV screen that has been polished attracts dust, dry hair crackles (and may actually spark!) when brushed, and you may feel a shock when you touch the metal door-handle of a car on getting out after a journey in dry weather. All these are examples of insulated objects that have gained an electric charge by friction – that is, by being rubbed against other objects.

Insulators that are charged by friction will attract other objects. Some of the effects have been known for centuries. Greek scientists experimented with amber that was charged by rubbing it with fur. Our words *electron* and *electricity* come from the Greek word ηλεκτρον, which means amber. Today, electrostatics experiments are often carried out with plastic materials which are moisture-repellent and stay charged for longer.

Charging by friction can be hazardous. For example, a lorry which carries a bulk powder must be earthed before emptying its load. Otherwise, electric charge can build up on the tanker. This could lead to a spark from the tanker to earth, causing an explosion. Similarly, the pipes used for movement of highly flammable liquids (for example, petrol), are metal-clad. An aircraft moving through the air will also become charged. To prevent the first person touching the aircraft on landing from becoming seriously injured, the tyres are made to conduct, so that the aircraft loses its charge on touchdown.

There are two kinds of electric charge. Polythene becomes negatively charged when rubbed with wool, and cellulose acetate becomes positively charged, also when rubbed with wool. To understand why this happens, we need to go back to the model of the atom. Atoms consist of a positively charged nucleus with negatively charged electrons orbiting it. When the polythene is rubbed with wool, friction causes some electrons to be transferred from the wool to the polythene. So the polythene has a negative charge, and the wool is left with a positive charge. Cellulose acetate becomes positive because it loses some electrons to the wool when it is rubbed. Polythene and cellulose acetate are poorly conducting plastics, and the charges remain static on the surface.

Putting two charged polythene rods close to each other, or two charged acetate rods close to each other, shows that similar charges repel each other (Figure 5.1). Conversely, unlike charges attract. A charged polythene rod attracts a charged acetate rod.

Figure 5.1 Like charges repel

This is the basic law of force between electric charges:

Like charges repel, unlike charges attract.

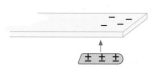

Figure 5.2 A charged rod can induce charges in an uncharged object

Charged rods will also attract uncharged objects. A charged polythene rod will pick up small pieces of paper, for example. The presence of the negative charge on the rod causes a redistribution of the charges in the paper. Free electrons are repelled to the far side of the paper, so that the side nearest the rod is positive and is therefore attracted to the rod (Figure 5.2). The paper is said to be **charged by induction**. When the rod is removed, the free electrons will move back and cancel the positive charge.

Force between charges

In the late eighteenth century, the French scientist Charles Coulomb investigated the magnitude of the force between charges, and how this varies with the charges involved and the distance between them. He discovered the following rule.

> The force is proportional to the product of the charges and inversely proportional to the square of the distance between them. This is known as **Coulomb's law**.

Coulomb's experiments made use of small, charged insulated spheres. Strictly speaking, the law applies to point charges, but it can be used for charged spheres provided that their radii are small compared with their separation.

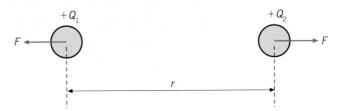

Figure 5.3

For point charges Q_1 and Q_2 a distance r apart (Figure 5.3), Coulomb's law gives the force F as

$$F \propto Q_1 Q_2 / r^2$$

or

$$F = \frac{kQ_1Q_2}{r^2}$$

where k is a constant of proportionality, the value of which depends on the insulating medium around the charges and the system of units employed. In SI units, F is measured in newtons, Q in coulombs and r in metres. Then the constant k is given as

$$k = \frac{1}{4\pi\epsilon_0}$$

and so

$$F = \frac{Q_1 Q_2}{4\pi\epsilon_0 r^2}$$

when the charges are in a vacuum. The quantity ϵ_0 is called the **permittivity of free space** (or the permittivity of a vacuum).

The value of the permittivity of air is very close to that of a vacuum ($1.0005\epsilon_0$), so the equation can be used for the force between charges in a vacuum or in air.

The value of the permittivity of free space is given by

$$\epsilon_0 = 8.85 \times 10^{-12}\,C^2\,N^{-1}m^{-2}$$

(We shall meet an alternative unit for ϵ_0 later, that is F m^{-1}.) This numerical value for ϵ_0 means that

$$k = \frac{1}{4\pi\epsilon_0} = 8.99 \times 10^9\,C^{-2}\,N\,m^2$$

Example

Calculate the force between two point charges, each of 1.0×10^{-9} C, which are 4.0 cm apart in a vacuum.

Using $F = kQ_1Q_2/r^2$, $F = 9.0 \times 10^9 \times (1.0 \times 10^{-9})^2/(4.0 \times 10^{-2})^2$
$$= \mathbf{5.6 \times 10^{-6}N}$$

Now it's your turn

Calculate the force between two electrons which are 1.0×10^{-10} m apart in a vacuum.
(electron charge $e = -1.6 \times 10^{-19}$ C)

Like Newton's law of gravitation, Coulomb's law is often referred to as an **inverse square** law of force, because the variation of force with distance r between the charges or masses is proportional to $1/r^2$. We say that the equations are analogous, or that there is an analogy between this aspect of electric forces and gravitational forces. Another analogy is that in each case the force is proportional to the product of the interacting charges or masses. However, there are important differences.

■ The electric force always acts on charges, whereas the gravitational force acts on masses.
■ The electric force can be attractive or repulsive, depending on the sign of the interacting charges, whereas two masses always attract each other.

Electric fields

Electric charges exert forces on each other when they are a distance apart. The idea of an **electric field** is used to explain this action at a distance. An electric field is a region of space where a stationary charge experiences a force.

Electric fields are invisible, but they can be represented by electric lines of force, just as magnetic fields can be represented by magnetic lines of force (see Chapter 12). The direction of the electric field is defined as the direction in which a positive charge would move if it were free to do so. So the lines of force can be drawn with arrows that go from positive to negative.

For any electric field:

- the lines of force start on a positive charge, and end on a negative charge
- the lines of force are smooth curves which never touch or cross
- the strength of the electric field is indicated by the closeness of the lines: the closer they are, the stronger the field.

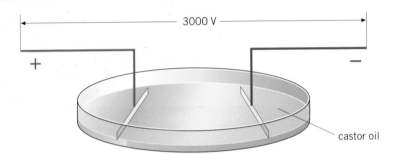

Figure 5.4 Apparatus for investigating electric field patterns

The apparatus in Figure 5.4 can be used to show electric field patterns. Semolina is sprinkled on a non-conducting oil and a high voltage supply is connected to the conducting plates. The semolina becomes charged by induction and lines up along the lines of force. Some electric field patterns are illustrated in Figure 5.5. The pattern for the charged conducting sphere (Figure 5.5c) is of particular importance. Although there are no electric field lines inside the conductor, the field lines appear to come from the centre. Thus, the charge on the surface of a conducting sphere appears as if it is all concentrated at the centre. This means that small conducting spheres may be used as an approximation to point charges.

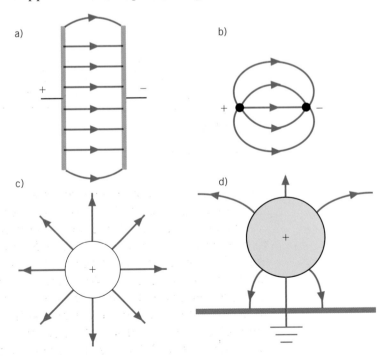

Figure 5.5 Some electric field patterns

Paint spraying makes use of some of the principles of electrostatics. It can be wasteful if the paint is not sprayed where it is needed. But if the spray is given a charge, and the object to be painted is given an opposite charge, the charged droplets follow the lines of force and end up on the surface. This is illustrated in Figure 5.6.

Figure 5.6 Paint spraying

Induced fields and charges

We have already stated that there is no resultant electric field in a conductor (unless a source of e.m.f. is connected across it). The reason for this is that electrons are free to move in the conductor. As soon as a charged body is placed near the conductor, the electric field of the charged body causes electrons in the conductor to move in the opposite direction to the field (because electrons have a negative charge). This is illustrated in Figure 5.7. These charges create a field in the opposite direction to the field due to the charged body. The induced charges will stop moving when the two fields are equal and opposite. Hence there is no resultant field in a conductor.

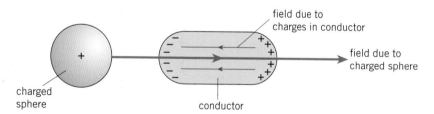

Figure 5.7 Induced charges

This effect may be used to charge a body by induction. The process is illustrated in Figure 5.8. If a positively charged rod is placed near an insulated conductor, induced charges appear on the conductor, as shown in Figure 5.8a. If the conductor is now earthed, as in 5.8b, electrons move from earth to neutralise the induced positive charge. The electrons are held in position by the positively charged rod. When the earth connection is removed, the negative charge is still held in position by the positively charged rod, as in 5.8c. Removal of the charged rod means that the electrons will distribute themselves over the surface of the sphere, as in 5.8d. Note that if a positively charged rod is used, the final charge on the sphere is negative; if a negatively charged rod is used, the sphere becomes positively charged.

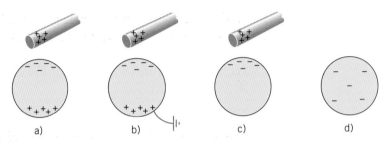

a)   b)   c)   d)

Figure 5.8 Charging by induction

Electric field strength

> The **electric field strength** at a point is defined as the force per unit charge acting on a small positive charge placed at that point.

If the force experienced by a positive test charge $+Q$ placed in the field is F, the field strength E is given by

$$E = \frac{F}{Q}$$

(Be careful not to be confused by the symbol E for field strength. E may also be used for energy.) A unit of field strength can be deduced from this equation: force is measured in newtons and charge in coulombs, so the SI unit of field strength is N C^{-1}. We shall see later that there is another common SI unit for electric field strength, V m^{-1}. These two units are equivalent.

A uniform electric field

In Figure 5.5a, the electric field pattern between the parallel plates consists of parallel, equally spaced lines, except near the edges of the plates. This means that the field between charged parallel plates (as, for example, in a capacitor) is uniform.

Figure 5.9 illustrates parallel plates a distance d apart with a potential difference V between them. A charge $+Q$ in the uniform field between the plates has a force F acting on it. To move the charge towards the positive plate would require work to be done on the charge. Work is given by the product of force and distance. To move the charge from one plate to the other requires work W given by

$$W = Fd$$

From the definition of potential difference (page 114 of the AS book),

$$W = VQ$$

Thus $W = Fd = VQ$, or, rearranging,

$$F/Q = V/d$$

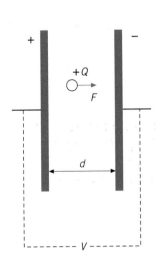

Figure 5.9

But F/Q is the force per unit charge, and this is the definition of electric field strength. Thus, for a uniform field, the field strength E is given by

$$E = \frac{V}{d}$$

This equation gives an alternative unit for electric field strength (remember that we have already derived N C^{-1} from the idea of force per unit charge). If we use the expression for a uniform field, the unit of potential difference is the volt and the unit of distance is the metre, so another unit for electric field strength is volts per metre (V m^{-1}). The two units N C^{-1} and V m^{-1} are equivalent.

Example

Two metal plates 5.0 cm apart have a potential difference of 1000 V between them. Calculate:
(a) the strength of the electric field between the plates,
(b) the force on a charge of 5.0 nC between the plates.

(a) From $E = V/d$, $E = 1000/5.0 \times 10^{-2} = \mathbf{2.0 \times 10^4 V\ m^{-1}}$.
(b) From $F = EQ$, $F = 2.0 \times 10^4 \times 5.0 \times 10^{-9} = \mathbf{1.0 \times 10^{-4} N}$.

Now it's your turn

Two metal plates 15 mm apart have a potential difference of 750 V between them. The force on a small charged sphere placed between the plates is 1.2×10^{-7} N. Calculate:
(a) the strength of the electric field between the plates,
(b) the charge on the sphere.

Field strength due to an isolated point charge

The electric field due to an isolated point charge is **radial** (see Figure 5.5c). We have to mention that the point charge is isolated. If any other object, charged or otherwise, is near it, the field would be distorted.

From Coulomb's law, the force on a test charge q a distance r from the isolated point charge Q is given by

$$F = \frac{Qq}{4\pi\epsilon_0 r^2}$$

The electric field E at the location of the test charge q is given by $E = F/q$. Thus, the electric field due to the isolated point charge is

$$E = \frac{Q}{4\pi\epsilon_0 r^2}$$

or

$$E = \frac{kQ}{r^2} \quad \text{where} \quad k = \frac{1}{4\pi\epsilon_0}$$

This is an inverse square law field of force because the force varies in proportion to $1/r^2$.

Example

In a simplified model of the hydrogen atom, the electron is at a distance of 5.3×10^{-11} m from the proton. The proton charge is $+1.6 \times 10^{-19}$ C. Calculate the electric field strength of the proton at this distance.

Assuming that the field is radial,
$E = kQ/r^2 = 9.0 \times 10^9 \times 1.6 \times 10^{-19}/(5.3 \times 10^{-11})^2$
$\qquad\qquad = \mathbf{5.1 \times 10^{11} N\ C^{-1}}$

Now it's your turn

A Van de Graaff generator has a dome of radius 15 cm. The dome carries a charge of 2.5×10^{-6} C. It can be assumed that this charge acts as if it were all concentrated at the centre of the spherical dome. Calculate the electric field strength at the dome's surface.

Because the electric field strength due to an isolated point charge is given by the force acting on a unit charge, and because this force obeys Coulomb's law, there is an analogy between the electric field due to a point charge and the gravitational field due to a point mass. This analogy shows exactly the same similarities and differences as were summarised on page 85.

Effect of an electric field on the motion of charged particles

An electric field exerts a force on a charged particle. This means that a positively charged particle will accelerate in the direction of the field. Its precise direction is given by the direction of the field line at the location of the charge. Similarly, a negatively charged particle accelerates in the opposite direction to the direction of the field. For example, in the uniform electric field between two parallel plates with a potential difference between them, the direction of the field is from the positive plate to the negative plate. An electron, because of its negative charge, will accelerate towards the positive plate. We can calculate the change in kinetic energy of the electron by equating the change in kinetic energy with the work done by the electric field in accelerating the charged particle. If the plates have a potential difference of V between them, an electron starting from rest from the negative plate will have a kinetic energy of $\frac{1}{2}mv^2$ when it reaches the positive plate given by

$$\frac{1}{2}mv^2 = eV.$$

You will find this equation very useful in calculating the speed of particles after acceleration through a potential difference.

If we have a beam of charged particles, the application of an electric field along the beam will cause the particles in the beam to accelerate or slow down, depending on the charge of the particles and the direction of the beam. If the field is applied at an angle to the beam, both speed and direction of the particles will be changed. Usually, we will deal with a uniform electric field applied at right angles to the original direction of the beam. An electric field has no effect on uncharged particles, of course. The details of the effects achieved will be discussed in Chapter 6, together with the way in which a magnetic field can be used to change the direction of the beam.

Example

Two metal plates have a potential difference of 2000 V between them. An electron is emitted from rest from the negative plate. What is its kinetic energy when it reaches the positive plate?
The electron charge is -1.6×10^{-19} C.
From $\frac{1}{2}mv^2 = eV$,

kinetic energy $= 1.6 \times 10^{-19} \times 2000 = \mathbf{3.2 \times 10^{-16}\ J}$

Now it's your turn

It is required to accelerate an electron from rest to a speed of $3.0 \times 10^6 \,\mathrm{m\,s^{-1}}$ by the application of a suitable potential difference. Calculate this potential difference.

Section 5.1 Summary

▪ Insulators may be charged by friction.
▪ Like charges repel; unlike charges attract each other.
▪ When charged objects are placed near conductors, they cause a redistribution of charge in the conductor, thereby inducing charges.
▪ The force between two point charges is proportional to the product of the charges and inversely proportional to the square of the distance between them. This is Coulomb's law:
$F = Q_1 Q_2/4\pi\epsilon_0 r^2$
▪ ϵ_0 is the permittivity of free space; its value is $8.85 \times 10^{-12}\ \mathrm{C^2 N^{-1} m^{-2}}$ $(\mathrm{F\,m^{-1}})$.
▪ An electric field is a region of space where a stationary charge experiences a force.

- Electric field strength is the force per unit positive charge: $E = F/Q$
- The electric field between parallel plates is uniform. The field strength is given by: $E = V/d$
- The electric field strength at a point in the field of an isolated point charge is given by: $E = Q/4\pi\epsilon_0 r^2$
- There is an analogy between electric and gravitational fields because both obey an inverse square law, but the direction of an electric field depends on the sign of the charge producing the field.
- Electric fields may be used to accelerate charged particles and to deflect beams of charged particles.

Section 5.1 Questions

1 A glass rod rubbed with silk becomes positively charged. Explain what has happened to the glass. Explain also why the charged rod is able to attract small pieces of paper.

2 (a) Explain what is meant by an electric field.
  (b) Sketch the electric field patterns
    (i) between two negatively charged particles,
    (ii) between a positive charge and a negatively charged flat metal plate.

3 In a simplified model, a uranium nucleus is a sphere of radius 8.0×10^{-15} m. The nucleus contains 92 protons (and rather more neutrons). The charge on a proton is 1.6×10^{-19} C. It can be assumed that the charge of these protons acts as if it were all concentrated at the centre of the nucleus. The nucleus releases an α-particle containing two protons (and two neutrons) at the surface of the nucleus. Calculate:
  (a) the electric field strength at the surface of the nucleus before emission of the α-particle,
  (b) the electric force on the α-particle at the surface of the nucleus.

5.2 Capacitance

At the end of Section 5.2 you should be able to:
- define capacitance and the farad
- recall and solve problems using $C = Q/V$
- deduce from the area under a potential–charge graph the equation $W = \frac{1}{2}QV$ and hence $W = \frac{1}{2}CV^2$
- derive, using the formula $C = Q/V$, conservation of charge and the addition of p.d.s, formulae for capacitors in series and in parallel
- solve problems using formulae for capacitors in series and in parallel
- sketch graphs that show the variation with time of potential difference, charge and current for a capacitor discharging through a resistor

▶▶

- ▨ define the time constant of a circuit
- ▨ recall and use the equation $\tau = CR$, where τ is the time constant
- ▨ analyse the discharge of a capacitor through a resistor using equations for exponential decay of the form $x = x_0 \exp(-t/CR)$, recognising this equation as a decay with a constant-ratio property
- ▨ show an understanding of the function of capacitors in simple circuits, and their application for the storage of energy.

Consider an isolated spherical conductor connected to a high voltage supply (Figure 5.10). it is found that, as the potential of the sphere is increased, the charge stored on the sphere also increases. The graph showing the variation of charge on the conductor with potential is shown in Figure 5.11. It can be seen that charge Q is related to potential V by

$$Q \propto V$$

Hence,

$$Q = CV$$

where C is a constant which depends on the size of the conductor. C is known as the capacitance of the conductor.

Capacitance is the ratio of charge to potential for a conductor.

$$C = \frac{Q}{V}$$

Another chance for confusion! The letter C is used as an abbreviation for the unit of charge, the coulomb; as an italic letter C it is used as the symbol for capacitance.

The unit of capacitance is the **farad** (symbol F). One farad is one coulomb per volt. The farad is an inconveniently large unit. In electronic circuits and laboratory experiments, the range of useful values of capacitance is from about 10^{-12} F (1 picofarad, or 1 pF) to 10^{-3} F (1 millifarad, or 1 mF).

Circuit components which store charge and therefore have capacitance are called **capacitors**.

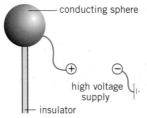

Figure 5.10 Charged spherical conductor

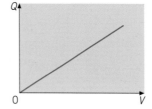

Figure 5.11 Relation between charge and potential

Calculate the charge stored on an isolated conductor of capacitance $470\,\mu F$ when it is at a potential of $20\,V$.

Using $C = Q/V$, we have $Q = CV = 470 \times 10^{-6} \times 20 =$ **$9.4 \times 10^{-3}\,C$**.

Now it's your turn

The charge on a certain isolated conductor is $2.5 \times 10^{-2}\,C$ when its potential is $25\,V$. Calculate its capacitance.

Energy stored in a capacitor

When charging a capacitor, work is done by the battery to move charge on to the capacitor. Energy is transferred from the power supply and is stored as **electric potential energy** in the capacitor.

Camera flash units use a capacitor to store energy. The capacitor takes a few seconds to charge when connected to the battery in the camera. Then the energy is discharged very rapidly when the capacitor is connected to the flash-bulb to give a short but intense flash.

The flash units used for optical stimulation of high-power lasers use a bank of capacitors to store the energy to activate what is essentially a very high intensity flash gun. Light of the correct frequency from the flash excites electrons from the ground state energy level of the lasing medium to a higher energy state, producing a situation of population inversion in the medium. The process is called optical pumping. The situation of population inversion is necessary if light amplification by the stimulated emission of radiation (laser action) is to take place. In both camera and laser flash guns the transfer of energy through the discharge tube takes place very rapidly.

Another application, this time in which it is desirable that the discharge of energy from the charged capacitor should take place very slowly, is in computers, in which the capacitor acts as a back-up power supply, for example to maintain the operation of the clock and calendar while the mains supply is disconnected.

Since $Q = CV$, the charge stored on a capacitor is directly proportional to its potential (see Figure 5.11).

From the definition of potential, the work done (and therefore the energy transferred) is the product of the potential and the charge. That is,

$$W \text{ (and } E_p\text{)} = VQ$$

However, while more and more charge is transferred to the capacitor, the potential difference is increasing. Suppose the potential is V_0 when the charge stored is Q_0. When a further small amount of charge Δq is supplied, the energy transferred is given by

$$\Delta E_p = V_0 \Delta q$$

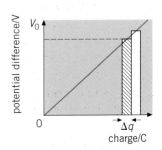

Figure 5.12 Graph of potential against charge for a capacitor

which is equal to the area of the shaded strip in Figure 5.12. Similarly, the energy transferred when a further charge Δq is transferred is given by the area of the next strip, and so on. If the amount of charge Δq is very small, the strips become very thin and their combined areas are just equal to the area between the line and the horizontal axis. Thus

> The energy transferred from the battery when a capacitor is charged is given by the area under the graph line when charge (*x*-axis) is plotted against potential difference (*y*-axis).

Because the graph is a straight line through the origin, this area is just the area of the right-angled triangle formed by the line and the charge axis. Thus

$$E_p = \tfrac{1}{2}QV$$

This is the expression for the energy transferred from the battery in charging the capacitor. This is electric potential energy, and it is released when the capacitor is discharged. Since $C = Q/V$, this expression can be written in different forms.

$$E_p = \tfrac{1}{2}QV = \tfrac{1}{2}CV^2 = \frac{Q^2}{2C}$$

Example

Calculate the energy stored by a 470 μF capacitor at a potential of 20 V.

Using $E_p = \tfrac{1}{2}CV^2$, $E_p = \tfrac{1}{2} \times 470 \times 10^{-6} \times (20)^2 = \textbf{0.094 J}$.

Now it's your turn

A camera flash-lamp uses a 5000 μF capacitor which is charged by a 9 V battery. Calculate the energy transferred when the capacitor is fully discharged through the lamp.

Capacitors

The simplest capacitor in an electric circuit consists of two metal plates, with an air gap between them which acts as an insulator. This is called a parallel-plate capacitor. Figure 5.13a shows the circuit symbol for a capacitor. When the plates are connected to a battery, the battery transfers electrons from the plate connected to the positive terminal of the battery to the plate connected to the negative terminal. Thus the plates carry equal but opposite charges.

The **capacitance** of a capacitor is defined as the charge stored on one plate per unit potential difference between the plates.

The capacitance of an air-filled capacitor can be increased by putting an insulating material, such as mica or waxed paper, between the plates. The material between the plates is called the **dielectric**. In a type of capacitor

a)

b)

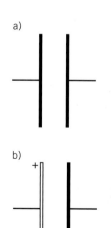

Figure 5.13 Circuit symbols for a) a capacitor and b) an electrolytic capacitor

known as an **electrolytic** capacitor the dielectric is deposited by an electrochemical reaction. These capacitors *must* be connected with the correct polarity for their plates, or they will be damaged. The circuit symbol for an electrolytic capacitor is shown in Figure 5.13b. Electrolytic capacitors are available with capacitances up to about 1 mF.

Capacitors in series and parallel

In Figure 5.14, the two capacitors of capacitance C_1 and C_2 are connected in series. We shall show that the combined capacitance C is given by

$$\frac{1}{C} = \frac{1}{C_1} + \frac{1}{C_2}$$

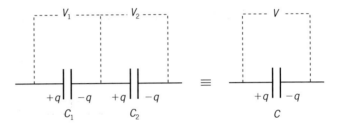

Figure 5.14 Capacitors in series

If the voltage across the equivalent capacitor is V and the charge stored on each plate is q, then $V = q/C$. The potential difference across the combination is the sum of the potential differences across the individual capacitors, $V = V_1 + V_2$, and each capacitor has charge q on each plate. A charge of $+q$ induced on one plate of one capacitor will induce a charge of $-q$ on the plate of the second capacitor. Since $V_1 = q/C_1$ and $V_2 = q/C_2$, then

$q/C = q/C_1 + q/C_2$

Dividing each side of the equation by q, we have

$1/C = 1/C_1 + 1/C_2$

A similar result applies for any number of capacitors connected in series.

The reciprocal of the combined capacitance equals the sum of the reciprocals of the individual capacitances in series.

Note that:

■ For two identical capacitors in series, the combined capacitance is equal to half of the value of each one.
■ For capacitors in series, the combined capacitance is always less than the value of the smallest individual capacitance.

In Figure 5.15, the two capacitors of capacitance C_1 and C_2 are connected in parallel. We shall show that the combined capacitance C is given by

$$C = C_1 + C_2$$

If the voltage across the equivalent capacitor is V and the charge stored on each plate is q, then $q = CV$. The total charge stored is the sum of the charges on the individual capacitors, $q = q_1 + q_2$, and there is the same potential difference V across each capacitor since they are connected in parallel. Since $q_1 = C_1V$ and $q_2 = C_2V$, then

$$CV = C_1V + C_2V$$

Dividing each side of the equation by V, we have

$$C = C_1 + C_2$$

The same result applies for any number of capacitors connected in parallel.

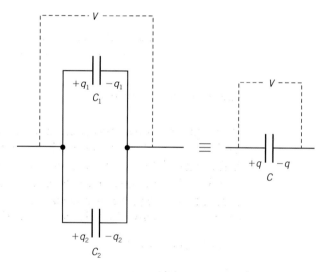

Figure 5.15 Capacitors in parallel

The combined capacitance equals the sum of all the individual capacitances in parallel.

Note that the equation for capacitors in *series* is similar to the equation for resistors in *parallel*, and the equation for capacitors in parallel is similar to the equation for resistors in series (see Chapter 5 of the AS book).

For the examination, you will not be required to derive the equations for the combined capacitance of capacitors in series and in parallel. We have worked through the derivations here because they provide examples of the way in which the ideas of the addition of potential differences, and of the principle of conservation of charge, can be applied. However, you will be expected to remember the equations. The trick is that they are the opposite way round from the equations for the combination of resistances.

Examples

1 A 100 μF capacitor in parallel with a 50 μF capacitor is connected to a 12 V supply. Calculate:
  (a) the total capacitance,
  (b) the potential difference across each capacitor,
  (c) the charge stored on each capacitor.

  (a) Using the equation for capacitors in parallel,
    $C = C_1 + C_2 = 100 + 50 = $ **150 μF.**
  (b) The potential difference across each capacitor is the same as the potential difference across the supply. This is **12 V.**
  (c) Using $Q = CV$, the charge stored on the 100 μF capacitor is $100 \times 10^{-6} \times 12 = $ **1.2×10^{-3} C.** The charge stored on the 50 μF capacitor is $50 \times 10^{-6} \times 12 = $ **6.0×10^{-4} C.**

2 A 100 μF capacitor in series with a 50 μF capacitor is connected to a 12 V supply. Calculate:
  (a) the combined capacitance,
  (b) the charge stored by each capacitor,
  (c) the potential difference across each capacitor.

  (a) Using the equation for capacitors in series,
    $1/C = 1/C_1 + 1/C_2 = 1/100 \times 10^{-6} + 1/50 \times 10^{-6} = 3 \times 10^4$.
    Thus $C = 3.3 \times 10^{-5} = $ **33 μF.**
  (b) The charge stored by each capacitor is the same as the charge stored by the combination, so
    $Q = CV = 33 \times 10^{-6} \times 12 = $ **4.0×10^{-4} C.**
  (c) Using $V = Q/C$, the potential difference across the 100 μF capacitor is $4.0 \times 10^{-4}/100 \times 10^{-6} = $ **4.0 V.** The potential difference across the 50 μF capacitor is $4.0 \times 10^{-4}/50 \times 10^{-6} = $ **8.0 V.** Note that the two potential differences add up to the supply voltage.

Now it's your turn

(a) A 470 μF capacitor is connected to a 20 V supply. Calculate the charge stored on the capacitor.
(b) The capacitor in (a) is then disconnected from the supply and connected to an uncharged 470 μF capacitor.
  (i) Explain why the capacitors are in parallel, rather than series, and why the total charge stored by the combination must be the same as the answer to (a).
  (ii) Calculate the capacitance of the combination.
  (iii) Calculate the potential difference across each capacitor.
  (iv) Calculate the charge stored on each capacitor.

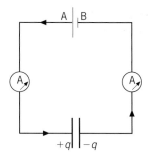

Figure 5.16 Charging a capacitor

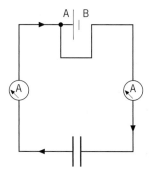

Figure 5.17 Discharging a capacitor

Charging and discharging capacitors

The circuit in Figure 5.16 can be used to investigate what happens when a capacitor is connected to a battery. Start with the lead to battery terminal A connected, but that to B disconnected. When connection is first made to terminal B, both ammeters flick to the right and then return to zero, indicating a momentary pulse of current. If the process is repeated, nothing further happens. The capacitor has become charged.

During the initial connection, the ammeters have similar deflections. This tells us that the same charge has moved on to the left-hand plate as has been removed from the right-hand plate. The momentary current has caused the left-hand plate to acquire a charge of $+q$, while the right-hand plate has a charge of $-q$. The capacitor is now said to store a charge of q (although if we add up the charges on the left- and right-hand plates taking account of their sign, the net charge is zero). There can be no steady current because of the gap between the capacitor plates. The current stops when the potential difference across the capacitor is the same as the e.m.f. of the battery.

When the battery lead to terminal B is disconnected and joined to point A (Figure 5.17) so that the battery is no longer in circuit, both ammeters give similar momentary deflections to the left. This time a current in the opposite direction has moved the charge of $+q$ from the left-hand plate to cancel the charge of $-q$ on the right-hand plate. The capacitor has become discharged.

Remember that in metal wires the current is carried by free electrons. These move in the opposite direction to that of the conventional current (see Chapter 5 of the AS book). When the capacitor is charged, electrons move from the negative terminal of the battery to the right-hand plate of the capacitor, and from the left-hand plate to the positive terminal of the battery. When the capacitor is discharged, electrons flow from the negative right-hand plate of the capacitor to the positive left-hand plate.

The experiment described using the circuit in Figure 5.17 showed that there is a momentary current when a capacitor discharges. A resistor connected in series with the capacitor will reduce the current, so that the capacitor discharges more slowly.

The circuit shown in Figure 5.18 can be used to investigate more precisely how a capacitor discharges. When the two-way switch is connected to point A, the capacitor will charge up until the potential difference between its plates is equal to the e.m.f. V_0 of the supply. When the switch is moved to B, the capacitor

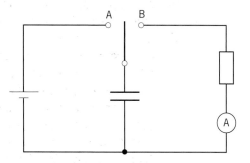

Figure 5.18 Circuit for investigating capacitor discharge

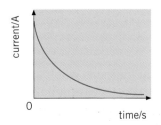

Figure 5.19 Graph of current against time for capacitor discharge

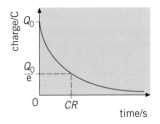

Figure 5.20 Graph of charge against time for capacitor discharge

will discharge through the resistor. When the switch makes contact with B, the current can be recorded at regular intervals of time as the capacitor discharges. A graph of the discharge current against time is shown in Figure 5.19. The current is seen to change rapidly at first, and then more slowly. More detailed analysis shows that the curve is **exponential**. We shall meet exponential changes again when we deal with the decay of radioactive substances (Chapter 7).

All exponential decay curves have an equation of the form

$$x = x_0 e^{-kt}$$

where x is the quantity that is decaying (and x_0 is the value of x at $t = 0$), e to three decimal places is the number 2.718 (the root of natural logarithms) and k is a constant characteristic of the decay. A large value of k means that the decay is rapid, and a small value means a slow decay.

The solution for the discharge of a capacitor of capacitance C through a resistor of resistance R is of the form

$$Q = Q_0 e^{-t/CR}$$

This is the equation of the graph in Figure 5.20. Here Q_0 is the charge on the plates at time $t = 0$, and Q is the charge at time t.

The graphs of Figures 5.19 and 5.20 have exactly the same shape, and thus the equation for the discharge of a capacitor may be written

$$I = I_0 e^{-t/CR}$$

Furthermore, since for a capacitor Q is proportional to V, then the solution for the discharge can be written as

$$V = V_0 e^{-t/CR}$$

Time constant

As time progresses, the exponential curve in Figure 5.20 gets closer and closer to the time axis, but never actually meets it. Thus, it is not possible to quote a time for the capacitor to discharge completely.

However, the quantity CR in the decay equation may be used to give an indication of whether the decay is fast or slow, as shown in Figure 5.21.

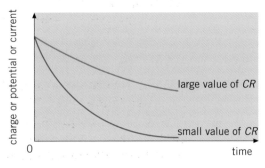

Figure 5.21 Decay curves for large and small time constants

> CR is called the **time constant** of the capacitor–resistor circuit.

CR has the units of time, and is measured in seconds.

We can easily show that CR has units of time. From $C = Q/V$ and $R = V/I$, then $CR = Q/I$. Since charge Q is in coulombs and current I is in amperes, and one ampere is equal to one coulomb per second, Q/I is in seconds.

To find the charge Q on the capacitor plates after a time $t = CR$, we substitute in the exponential decay equation

$$Q = Q_0 e^{-CR/CR} = Q_0 e^{-1} = Q_0/e = Q_0/2.718$$

Thus

> The time constant is the time for the charge to have decreased to $1/e$ (or $1/2.718$) of its initial charge.

In one time constant the charge stored by the capacitor drops to roughly one-third of its initial value. During the next time constant it will drop by the same ratio, to about one-ninth of the value at the beginning of the decay. This constant-ratio decrease in equal time intervals is a characteristic of the exponential decay equation.

Example

A $500\,\mu\text{F}$ capacitor is connected to a $10\,\text{V}$ supply, and is then discharged through a $100\,\text{k}\Omega$ resistor. Calculate:
(a) the initial charge stored by the capacitor,
(b) the initial discharge current,
(c) the value of the time constant,
(d) the charge on the plates after $100\,\text{s}$,
(e) the time at which the remaining charge is $2.5 \times 10^{-3}\,\text{C}$.

(a) From $Q = CV$, we have $Q = 500 \times 10^{-6} \times 10 = \mathbf{5.0 \times 10^{-3}\,C}$.
(b) From $I = V/R$, we have $I = 10/(100 \times 10^3) = \mathbf{1.0 \times 10^{-4}\,A}$.
(c) $CR = 500 \times 10^{-6} \times 100 \times 10^3 = \mathbf{50\,s}$.
(d) After $50\,\text{s}$, the charge on the plates is
$Q_0/e = 5 \times 10^{-3}/2.718 = 1.8 \times 10^{-3}$ C; after another $50\,\text{s}$, the charge is $1.8 \times 10^{-3}/2.718 = \mathbf{6.8 \times 10^{-4}\,C}$.
(e) Using $Q = Q_0 e^{-t/CR}$, $2.5 \times 10^{-3} = 5.0 \times 10^{-3} e^{-t/50}$, or $0.50 = e^{-t/50}$. Taking natural logarithms of both sides, $-0.693 = -t/50$, or $t = \mathbf{35\,s}$.

Now it's your turn

A $5.0\,\mu\text{F}$ capacitor is charged from a $12\,\text{V}$ battery, and is then discharged through a $0.50\,\text{M}\Omega$ resistor. Calculate:
(a) the initial charge on the capacitor,
(b) the charge on the capacitor $2.0\,\text{s}$ after the discharge starts,
(c) the potential difference across the capacitor at this time.

Section 5.2 Summary

- A capacitor stores charge. Its capacitance C is given by $C = Q/V$, where Q is the charge on the capacitor when there is a potential difference V between its plates.
- The unit of capacitance, the farad (F), is one coulomb per volt.
- The energy stored in a charged capacitor is given by:
  $E = \frac{1}{2}QV = \frac{1}{2}CV^2 = \frac{1}{2}Q^2/C$
- The equivalent capacitance C of capacitors connected in series is given by: $1/C = 1/C_1 + 1/C_2 + \ldots$
- The equivalent capacitance C of capacitors connected in parallel is given by: $C = C_1 + C_2 + \ldots$
- When a charged capacitor discharges, the charge on the plates decays exponentially. The equation for the decay is: $Q = Q_0 e^{-t/CR}$
- The time constant of the circuit, given by CR, is the time for the charge to decay to $1/e$ of its initial value.

Section 5.2 Questions

1 A parallel-plate capacitor has rectangular plates of side 200 mm by 30 mm. The plates are separated by an air gap 0.50 mm thick. The permittivity of free space is $8.85 \times 10^{-12}\,\mathrm{F\,m^{-1}}$.
   (a) Calculate the capacitance of the capacitor.
   (b) The capacitor is connected to a 12 V battery. Calculate:
      (i) the charge on each plate,
      (ii) the electric field between the plates.

2 Figure 5.22 shows an arrangement of capacitors.

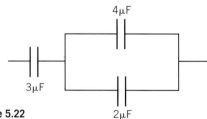

Figure 5.22

(a) Calculate the capacitance of this arrangement.
(b) The 4 μF capacitor is disconnected. Calculate the new capacitance.

3 A 10 μF capacitor is charged from a 9.0 V battery.
   (a) Calculate:
      (i) the electric potential energy stored by the capacitor,
      (ii) the charge stored by the capacitor.
   (b) The charged capacitor is discharged through a 150 kΩ resistor. Calculate:
      (i) the initial discharge current,
      (ii) the time constant,
      (iii) the time taken for the current to fall to 3.6×10^{-5} A.

Exam-style Questions

1 Two metal plates 10 cm apart are connected to a 1000 V supply. A small plastic ball with a conducting surface is suspended between the plates on an insulating thread. When the ball is made to touch the positive plate, it gains a charge of 5.0×10^{-9} C.

(a) Calculate the strength of the electric field between the plates.

(b) Explain why the ball shuttles to and fro between the plates.

(c) Calculate the force on the ball when it is in the field.

2 In a simplified model of the hydrogen atom, the electron is separated from the proton by a distance r of 5.3×10^{-11} m. Use the following data to calculate the ratio of the electric force to the gravitational force at this distance. What would be the ratio if the charges were separated to a distance $2r$?

Electron charge $e = -1.6 \times 10^{-19}$ C
Electron mass $m_e = 9.1 \times 10^{-31}$ kg
Proton mass $m_p = 1.7 \times 10^{-27}$ kg
Permittivity of free space $\epsilon_0 = 8.9 \times 10^{-12}$ F m^{-1}
Gravitational constant $G = 6.7 \times 10^{-11}$ N m^2 kg^{-2}

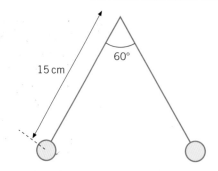

Figure 5.23

3 In Figure 5.23, there are two similar table-tennis balls, each of mass 2.5 g. The balls have equal charges and are suspended by insulating thread.

(a) Calculate the tension in one of the threads.

(b) Show that the electrostatic force between the two balls is 1.4×10^{-2} N.

(c) Calculate the charge on each ball.

6. Electromagnetism

A study of electricity and of electromagnetic effects is one of the key areas for all students of physics. Not only are electromagnetic effects vital to modern living (imagine life without the electric motor!) but they also give some insight into the pioneering research carried out by scientists such as Volta, Ampere, Ohm and Faraday. Their discoveries were exceptional because they were attempting to understand the invisible effects of electric currents and magnetic fields.

6.1 Magnetic fields

At the end of Section 6.1 you should be able to:
- show an understanding that a magnetic field is a field of force produced either by current-carrying conductors or by permanent magnets
- represent a magnetic field by means of field lines
- sketch flux patterns due to a long straight wire, a flat circular coil and a long solenoid
- show an understanding that the field due to a solenoid may be influenced by the presence of a ferrous core.

Some of the properties of magnets have been known for many centuries. The ancient Greeks discovered an iron ore called lodestone which, when hung from a thread, would come to rest always pointing in the same direction. This was the basis of the magnetic compass which has been in use since about 1500BC as a means of navigation.

The magnetic compass is dependent on the fact that a freely suspended magnet will come to rest pointing north–south. The ends of the magnet are said to be poles. The pole pointing to the north is referred to as the north-seeking pole (the north pole) and the other, the south-seeking pole (the south pole). It is now known that a compass behaves in this way because the Earth is itself a magnet.

Magnets exert forces on each other. These forces of either attraction or repulsion are used in many children's toys, in door catches and 'fridge magnets'. The effects of the forces may be summarised in the **law of magnets**.

Like poles repel. Unlike poles attract.

The law of magnets implies that around any magnet, there is a region where a magnetic pole will experience a force. This region is known as a **magnetic field**. Magnetic fields are not visible but they may be represented by lines of magnetic force or **magnetic field lines**, or magnetic 'flux'. A simple way of imaging magnetic field lines is to think of one such line as the direction in which a free magnetic north pole would move if placed in the field. Magnetic field lines may be plotted using a small compass (a plotting compass) or by the use of iron filings and a compass (Figure 6.1).

Figure 6.1 The iron filings line up with the magnetic field of the bar magnet which is under the sheet of paper. A plotting compass will give the direction of the field.

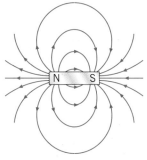

Figure 6.2 Magnetic field pattern of a bar magnet

The magnetic field lines of a bar magnet are shown in Figure 6.2. Effects due to the Earth's magnetic field have not been included since the Earth's field is relatively weak and would be of importance only at some distance from the magnet. It is important to realise that, although the magnetic field has been drawn in two dimensions, the actual magnetic field is three-dimensional.

For any magnetic field:

- the magnetic field lines start at a north pole and end at a south pole
- the magnetic field lines are smooth curves which never touch or cross
- the strength of the magnetic field is indicated by the distance between the lines – closer lines mean a stronger field.

It can be seen that these properties are very similar to those for electric field lines (see page 86).

Figure 6.3 illustrates the magnetic field pattern between the north pole of one magnet and the south pole of another. This pattern is similar to that produced between the poles of a horseshoe magnet.

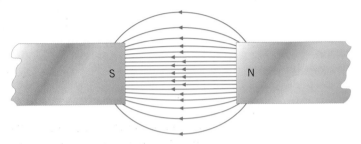

Figure 6.3 Magnetic field pattern between a north and south pole

Figure 6.4 shows the magnetic field between the north poles of two magnets. The magnetic field due to one magnet opposes that due to the other. The field lines cannot cross and consequently there is a point X, known as a **neutral point**, where there is no resultant magnetic field because the two fields are equal in magnitude but opposite in direction.

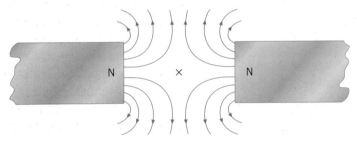

Figure 6.4 Magnetic field pattern between two north poles

Example

A circular magnet is made with its north pole at the centre, separated from the surrounding circular south pole by an air gap. Draw the magnetic field lines in the gap.

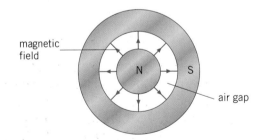

Figure 6.5

Now it's your turn

1 Draw a diagram to illustrate the magnetic field between the south poles of two magnets.

2 Two bar magnets are placed on a horizontal surface (Figure 6.6).

Figure 6.6

Draw the two magnets, and sketch the magnetic field lines around them. On your diagram, mark the position of any neutral points.

Magnetic effect of an electric current

Figure 6.7 Hans Christian Oersted

The earliest discovery of the magnetic effect of an electric current was made in 1820 when Oersted, a Danish physicist, noticed that a compass was deflected when brought near to a wire carrying an electric current. It is now known that all electric currents produce magnetic fields. The size and shape of the magnetic field depends on the size of the current and the shape (configuration) of the conductor through which the current is travelling.

The magnetic field due to a long straight wire may be plotted using the apparatus illustrated in Figure 6.8. Note that the current must be quite large (about 5 A). Iron filings are sprinkled on to the horizontal board and a plotting compass is used to determine the direction of the field.

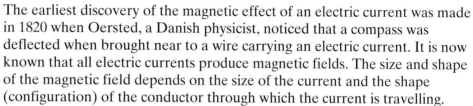

Figure 6.8 Apparatus to plot the magnetic field due to a long wire

Figure 6.9 shows the field pattern due to a long straight current-carrying wire. The lines are concentric circles centred on the middle of the wire. The separation of the lines increases with distance from the wire, indicating that the field is decreasing in strength. The direction of the magnetic field may be found using the **right-hand rule** (see Figure 6.10).

Imagine holding the conductor in the right hand with the thumb pointing in the direction of the current. The direction of the fingers gives the direction of the magnetic field.

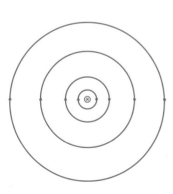

Figure 6.9 Magnetic field pattern due to a long straight wire

Similar apparatus to that in Figure 6.8 may be used to investigate the shapes of the magnetic field due to a flat coil and to a solenoid (a long coil).

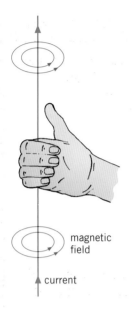

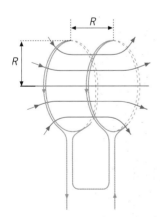

Figure 6.10 The right-hand rule

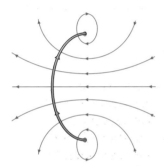

Figure 6.11 Magnetic field pattern due to a flat coil

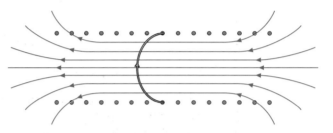

Figure 6.11 illustrates the magnetic field pattern due to a flat coil. The field has been drawn in a plane normal to the coil and through its centre.

A solenoid may be thought to be made up of many flat coils placed side-by-side. The magnetic field pattern of a long solenoid (that is, a coil which is long in comparison with its diameter) is shown in Figure 6.12. The field lines are parallel and equally spaced over the centre section of the solenoid, indicating that the field is uniform. The field lines spread out towards the ends.

Figure 6.12 Magnetic field pattern due to a solenoid

The strength of the magnetic field at each end is one half that at the centre. The direction of the magnetic field in a flat coil and in a solenoid may be found using the **right-hand grip rule**, as illustrated in Figure 6.13.

Grasp the coil or solenoid in the right hand with the fingers pointing in the direction of the current. The thumb gives the direction of the magnetic field.

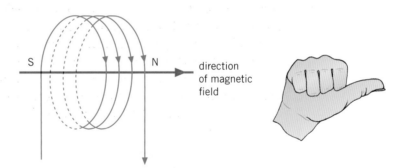

Figure 6.13 The right-hand grip rule

The magnetic north end of the coil or solenoid is the end from which the lines of magnetic force are emerging. Note the similarities and, more importantly, the differences between this rule and the right-hand rule for the long straight wire (Figure 6.10).

Uniform magnetic fields are of importance in the study of charged particles. A uniform field is produced in a solenoid but this field is inside the solenoid and consequently, it may be difficult to make observations and to take measurements. This problem is overcome by using **Helmholtz coils**. These are two identical flat coils, with the same current in each. The coils are positioned so that their planes are parallel and separated by a distance equal to the radius of either coil. The coils and their resultant magnetic field are illustrated in Figure 6.14.

Figure 6.14 Magnetic field in Helmholtz coils

Example

Two long straight wires, of circular cross-section, are each carrying the same current directly away from you, down into the page (Figure 6.15). Draw the magnetic field due to the two current-carrying wires.

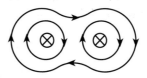

Figure 6.15

Now it's your turn

1 Draw magnetic field patterns, one in each case to represent
   (a) a uniform field,
   (b) a field which is decreasing in strength in the direction of the field,
   (c) a field which is increasing in strength along the direction of the field.

2 Draw a diagram of the magnetic field due to two long straight wires when the currents in the two wires are in opposite directions.

Electromagnets and their uses

The strength of the magnetic field due to a flat coil or a solenoid may be increased by winding the coil on a bar of soft iron. The bar is said to be the **core** of the coil. The iron is referred to as being 'soft' because it can be magnetised and demagnetised easily. With such a core, the strength of the magnetic field may be increased by up to 1000 times. With ferrous alloys (iron alloyed with cobalt or nickel), the field may be 10^4 times stronger. Magnets such as these are called **electromagnets**. Electromagnets have many uses because, unlike a permanent magnet, the magnetic field can be switched off by switching off the current in the coil.

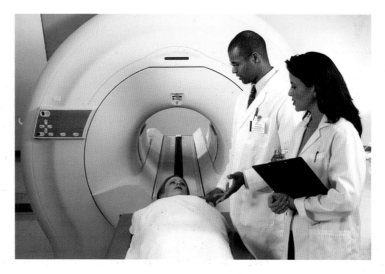

Figure 6.16 Magnetic imaging of body structures requires the use of a very large uniform magnetic field

Section 6.1 Summary

- A magnetic field is a region of space where a magnet will experience forces.
- Magnetic field lines never touch or cross.
- The separation of magnetic field lines indicates the strength of the magnetic field.
- An electric current gives rise to a magnetic field, the strength and direction of which depends on the size of the current and the shape of the current-carrying conductor.
- The direction of the field due to a straight wire is given by the right-hand rule.
- The direction of the field in a solenoid is given by the right-hand grip rule.
- The field of a solenoid may be increased in strength by a ferrous core; this is the principle of an electromagnet.

Section 6.1 Questions

1 A plotting compass may be used to investigate magnetic fields (see page 105). When used to investigate the magnetic field due to the Earth, it is often found that the compass needle is not always balanced parallel to the Earth's surface, but one end dips down. What do you deduce?

2 The current in domestic wiring produces an associated magnetic field. Discuss what might be observed if you tried to trace the wiring using a plotting compass.

6.2 Forces and fields

At the end of Section 6.2 you should be able to:

- show an appreciation that a force might act on a current-carrying conductor placed in a magnetic field
- recall and solve problems using the equation $F = BIl \sin \theta$, with directions as interpreted by Fleming's left-hand rule
- define magnetic flux density and the tesla
- show an understanding of how the force on a current-carrying conductor can be used to measure the flux density of a magnetic field, using a current balance
- explain the forces between current-carrying conductors and predict the direction of the forces
- predict the direction of the force on a charge moving in a magnetic field
- recall and solve problems using $F = Bqv \sin \theta$

- show an appreciation of the main principles of the determination of *e* by Millikan's experiment
- summarise and interpret the experimental evidence for quantisation of charge
- explain the main principles of one method for the determination of ν and e/m for electrons
- describe and analyse qualitatively the deflection of beams of charged particles by uniform electric and magnetic fields
- analyse quantitatively the circular orbits of charged particles moving in a plane perpendicular to a uniform magnetic field by relating the magnetic force to the centripetal acceleration it causes
- explain how electric and magnetic fields can be used in velocity selection
- explain the use of the deflection of charged particles in the magnetic and electric fields of a mass spectrometer.

In Section 6.1, we saw that the plotting of lines of magnetic force gives the direction and shape of the magnetic field. Also, the distance between the lines indicates the strength of the field. However, the strength of the magnetic field was not defined. Physics is the science of measurement and consequently, the topic would not be complete without defining and measuring magnetic field strength. Magnetic field strength is defined through a study of the **motor effect**.

The motor effect

The interaction of the magnetic fields produced by two magnets causes forces of attraction or repulsion between the two. If a conductor is placed between magnet poles and a current is passed through the conductor, the magnetic fields of the current-carrying conductor and the magnet may interact, causing forces between them. This is illustrated in Figure 6.17.

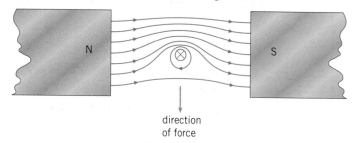

direction
of force

Figure 6.17 The interacting magnetic fields of a current-carrying conductor and the poles of a magnet

The existence of the force may be demonstrated with the apparatus shown in Figure 6.18. The strip of aluminium foil is held loosely between the poles of a horseshoe magnet so that the foil is at right-angles to the magnetic field. When the current is switched on, the foil jumps and becomes taut, showing that a force is acting on it. The direction of the force, known as the **electromagnetic force**, depends on the directions of the magnetic field and of the current. This phenomenon, when a current-carrying conductor is at an angle to a magnetic field, is called the **motor effect**.

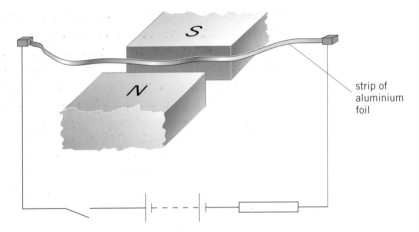

Figure 6.18 Demonstrating the motor effect acting on a piece of aluminium

The direction of the force relative to the directions of the current and the magnetic field may be predicted using **Fleming's left-hand rule**. This is illustrated in Figure 6.19.

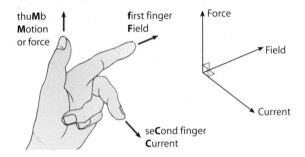

Figure 6.19 Fleming's left-hand rule

If the first two fingers and thumb of the left hand are held at right angles to one another with the **F**irst finger in the direction of the **F**ield and the se**C**ond finger in the direction of the **C**urrent, then the thu**M**b gives the direction of the force or **M**otion.

Note that, if the conductor is held fixed, motion will not be seen but nevertheless, there will be an electromagnetic force.

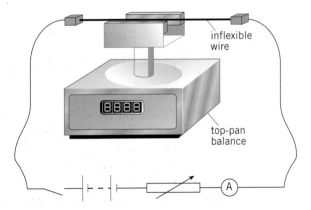

Figure 6.20 Apparatus to investigate the magnitude of the electromagnetic force

The magnitude of the electromagnetic force may be investigated using the 'current balance' apparatus shown in Figure 6.20. A horseshoe magnet is made by placing two flat magnadur magnets on a U-shaped iron core. A length of inflexible wire is firmly fixed between the poles of the magnet. When the current is switched on, the reading on the top-pan balance changes. This change in reading is a measure of the electromagnetic force. Variation of the current leads to the conclusion that the electromagnetic force F is proportional to the current I. By using magnadur magnets of different lengths (but of similar strengths), the force F is found to be proportional to the length of wire l in the magnetic field. Finally, by varying the angle θ between the wire and the direction of the magnetic field, the force F is found to be proportional to $\sin \theta$. Hence the expression

$$F \propto Il \sin \theta$$

is derived. The expression can be rewritten as

$$F = BIl \sin \theta$$

where B is a constant.

The constant B depends on the strength of the magnet and, if stronger magnets are used, the constant has a greater value. The equation can, therefore, be used as the defining equation for magnetic field strength.

The equation can be rewritten as

$$\frac{F}{l} = BI \sin \theta$$

So,

For a long straight conductor carrying unit current at right-angles to a uniform magnetic field, the magnetic field strength B is numerically equal to the force per unit length of the conductor.

Magnetic field strength is usually referred to as **magnetic flux density**. In SI units, this is measured in **tesla (T)**. An alternative name for this unit is **weber per square metre (Wb m^{-2})**.

One tesla is the uniform magnetic flux density which, acting normally to a long straight wire carrying a current of 1 ampere, causes a force per unit length of 1 N m^{-1} on the conductor.

Since force is measured in newtons, length in metres and current in amperes, it can be derived from the defining equation for the tesla that the tesla may also be expressed as N m^{-1} A^{-1}. The unit involves force which is a vector quantity and thus magnetic flux density is also a vector.

When using the equation $F = BIl \sin \theta$, it is sometimes helpful to think of the term $B \sin \theta$ as being the component of the magnetic flux density which is at right-angles (or normal) to the conductor (see Figure 6.21).

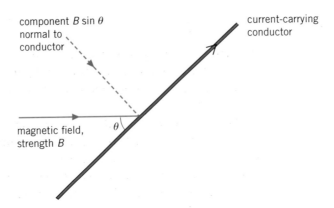

Figure 6.21 $B \sin \theta$ is the component of the magnetic field which is normal to the conductor

The tesla is a large unit for the measurement of flux density. A strong magnet may have a flux density between its poles of a few teslas. The magnetic flux density due to the Earth in the UK is about 44 μT at an angle of 66° to the horizontal.

Example

The horizontal component of the Earth's magnetic flux density is 1.8×10^{-5} T. The current in a horizontal cable is 150 A. Calculate, for this cable:
(a) the maximum force per unit length,
(b) the minimum force per unit length.
In each case, state the angle between the cable and the magnetic field.

(a) force per unit length $= \dfrac{F}{l} = BI \sin \theta$

Force per unit length is a maximum when $\theta = 90°$ and $\sin \theta = 1$.
Force per unit length $= 1.8 \times 10^{-5} \times 150 = 2.7 \times 10^{-3} \text{N m}^{-1}$
Maximum force per unit length = **2.7×10^{-3} N m^{-1}** when the cable is at right-angles to the field.

(b) Force per unit length is a minimum when $\theta = 0$ and $\sin \theta = 0$.
Minimum force per unit length = **0** when the cable is along the direction of the field.

Now it's your turn

1 The effective length of the filament in a light bulb is 3.1 cm and, for normal brightness, the current in the filament is 0.25 A. Calculate the maximum electromagnetic force on the filament when in the Earth's field of flux density 44 μT.

2 A straight conductor carrying a current of 6.5 A is situated in a uniform magnetic field of flux density 4.3 mT. Calculate the electromagnetic force per unit length of the conductor when the angle between the conductor and the field is:
(a) 90°,
(b) 45°.

Force between parallel conductors

A current-carrying conductor has a magnetic field around it. If a second current-carrying conductor is placed parallel to the first, this second conductor will be in the magnetic field of the first and, by the motor effect, will experience a force.

By similar reasoning, the first conductor will also experience a force. By Newton's third law (see Chapter 1), these two forces will be equal in magnitude and opposite in direction. The effect can be demonstrated using the apparatus in Figure 6.22. It can be seen that, if the currents are in the same direction, the pieces of foil move towards one another (the pinch effect). If the currents are in opposite directions, the pieces of foil move apart. An explanation for the effect can be found by reference to Figure 6.23.

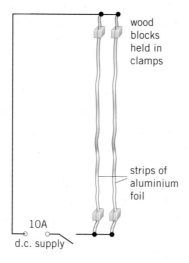

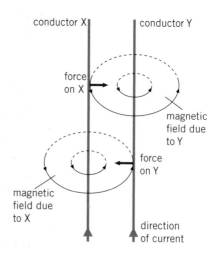

Figure 6.22 Apparatus to demonstrate the force between parallel current-carrying conductors

Figure 6.23 Diagram to illustrate the force between parallel current-carrying conductors

The current in conductor X causes a magnetic field and the field lines are concentric circles. These field lines will be at right angles to conductor Y and so, using Fleming's left-hand rule, there will be a force on Y in the direction of X. Using similar reasoning, the force on X due to the magnetic field of Y is towards Y. Reversing the direction of the current in one conductor will reverse the directions of the two forces and thus, when the currents are in opposite directions, the conductors tend to move apart.

The force per unit length on each of the conductors depends on the magnitude of the current in each conductor and also their separation. Since force can be expressed in terms of base units, then it is possible to use the effect to define the ampere in terms of SI units. The definition may be given as follows.

Consider two long straight parallel conductors of negligible area of cross-section, situated one metre apart in a vacuum. Then, if the current in each conductor is 1 ampere, the force per metre length acting on each conductor is $2.0 \times 10^{-7}\,\mathrm{N}$.

It may not be necessary to learn the definition of the ampere, but it is important to realise how the ampere is defined in terms of SI units and that the definition does not involve the property of a substance.

Force on a moving charged particle in a magnetic field

An electric current is charge in motion. Since charge is always associated with particles, then the current in a conductor is a movement of charged particles. If a current-carrying conductor is placed in a magnetic field, it may experience a force depending on the angle between the field and the conductor. The force arises from the force on the individual moving charged particles in the conductor. It has been shown that a conductor of length l carrying a current I at an angle θ to a uniform magnetic field of flux density B experiences a force F given by

$F = BIl \sin \theta$

If there are n charged particles in a length l of the conductor, each carrying a charge q, which pass a point in the conductor in time t, then the current in the conductor is given by

$$I = \frac{nq}{t}$$

and the speed of charged particles is given by $v = \dfrac{l}{t}$. Hence,

$$F = B\frac{nq}{t} l \sin \theta$$

and

$F = Bnqv \sin \theta$

This force is the force on n charged particles. Therefore,

> The force on a particle of charge q moving at speed v at an angle θ to a uniform magnetic field of flux density B is given by
>
> $F = Bqv \sin \theta$

The direction of the force will be given by Fleming's left-hand rule. However, it must be remembered that the second finger is used to indicate the direction of the *conventional* current. If the particles are positively charged, then the second finger is placed in the same direction as the velocity. However, if the particles are negatively charged (e.g. electrons), the finger must point in the opposite direction to the velocity.

Consider a positively charged particle of mass m carrying charge q and moving with velocity v as shown in Figure 6.24.

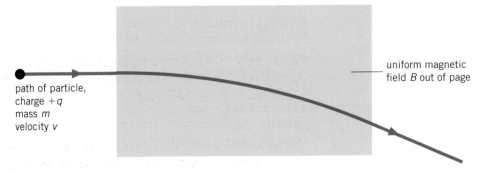

path of particle,
charge $+q$
mass m
velocity v

uniform magnetic
field B out of page

Figure 6.24

The particle enters a uniform magnetic field of flux density B which is normal to the direction of motion of the particle. As the particle enters the field, it will experience a force normal to its direction. This force will not change the speed of the particle but it will change its direction of motion. As the particle moves through the field, the force will remain constant since the speed has not changed and it will always be normal to the direction of motion. The particle will, therefore, move in an arc of a circle.

The force $F = Bqv \sin \theta$ (in this case, $\sin \theta = 1$), provides the centripetal force for the circular motion. Hence,

centripetal force $= mv^2/r = Bqv$

Re-arranging,

$$\text{radius of path} = r = \frac{mv}{Bq}$$

The importance of this equation is that, if the speed of the particle and the radius of its path are known, then the specific charge, i.e. the ratio of charge to mass, can be found. Then, if the charge on the particle is known, its mass may be calculated. The technique is also used in nuclear research to try to identify some of the fundamental particles. The tracks of charged particles are made visible in a bubble chamber (Figure 6.25). Analysing these tracks gives information as to the sign of the charge on the particle and its specific charge.

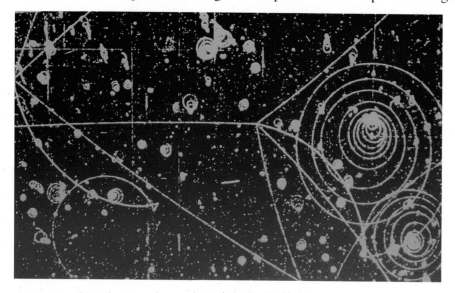

Figure 6.25 Tracks of particles produced in a bubble chamber

Specific charge of the electron

We will now look at two important experiments, both involving forces on charges in fields, which led to information on fundamental properties of the electron, its charge and its mass.

The electron charge

The principle of a method to measure the electron charge is very simple. It involves the balancing of the gravitational force on a tiny, charged sphere by an electric force in the opposite direction, so that the sphere remains

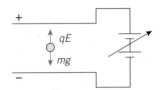

Figure 6.26 Principle of Millikan's experiment for e

stationary. In Millikan's original experiment, the sphere was an oil droplet, which became charged by friction as it passed out of the jet of a spray. If you do this experiment in a school laboratory now, you will probably use a very small polystyrene sphere.

Millikan's arrangement is sketched in Figure 6.26. The charged droplets are sprayed between two parallel plates connected to a source of variable potential difference, so that the electric field between them can be changed. The tendency is for a droplet to fall under its own weight, mg. The observer concentrates on one droplet and adjusts the electric field until the weight is exactly balanced by the electric force, qE, in the opposite direction. This gives the equilibrium condition $qE = qV/d = mg$. Millikan found that the charge q was always an integral multiple of a fundamental quantity of charge e equal to 1.6×10^{-19} C. This fundamental charge was ascribed to the charge of the electron. The experimental result that the charges on the droplets seems to be only in integral multiples of e means that charge is quantised, or exists only in discrete amounts.

Millikan's determination of the fundamental charge is regarded as one of the most beautiful experiments in physics. It makes use of the simplest possible physical principles, and yet the information that came out of it was of the utmost importance in establishing the value of one of the most necessary quantities in physics, the fundamental charge e. It also established the principle of the quantisation of charge, another vital fact. Millikan was awarded the 1923 Nobel Prize for this work.

You will notice that the forces involved in Millikan's experiment are electric and gravitational, and you may feel that this topic is out of place in a chapter on Electromagnetism. We include it here because of its context with another classic experiment on the properties of the electron, the determination of its specific charge, which involves the interaction of moving electrons with a magnetic field.

Example

A negatively charged polystyrene sphere of mass 3.3×10^{-15} kg is held at rest between two parallel plates separated by 5.0 mm when the potential difference between them is 170 V. How many excess electrons are on the sphere?

Apply the equation for the equilibrium condition, $qV/d = mg$. This gives $q = 3.3 \times 10^{-15} \times 9.8 \times 5.0 \times 10^{-3}/170 = 9.5 \times 10^{-19}$ C. Taking the electron charge e as 1.6×10^{-19} C, q corresponds to 5.9 excess electrons. Because you can't have 0.9 of an electron, the answer must be **6 excess electrons**. The discrepancy must be due to experimental inaccuracy or to rounding errors; we have been working to only two significant figures.

Now it's your turn

The charged sphere in the example above suddenly loses one of its excess electrons. The potential difference between the plates remains the same. Describe the initial motion of the sphere.

The electron mass

How do you find the mass of an electron? This cannot be done directly but the ratio of charge to mass e/m can be found and, knowing the charge e on the electron, the mass m can then be calculated.

We have seen that a particle of mass m and charge q moving with speed v at right angles to a uniform magnetic field of flux density B experiences a force F_B given by

$$F_B = Bqv$$

The direction of this force is given by Fleming's left-hand rule and is always normal to the velocity. The fact that we have a constant force which is normal to the direction of motion is the condition necessary for circular motion. The force F_B provides the centripetal force.

The motion can be represented by the equation

$$Bqv = mv^2/r$$

Re-arranging the terms,

$$q/m = v/Br$$

The ratio e/m for an electron, sometimes referred to as the specific charge, may be determined using a fine-beam tube, as shown in Figure 6.27. The path of electrons is made visible by having low-pressure gas in the tube and thus the radius of the orbit may be measured. By accelerating the electrons through a known potential difference, their speed on entry into the region of the magnetic field may be calculated (see Chapter 5 of the AS book). The magnetic field is provided by a pair of current-carrying coils (Helmholtz coils).

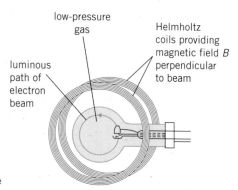

low-pressure gas

Helmholtz coils providing magnetic field B perpendicular to beam

luminous path of electron beam

Figure 6.27 Fine-beam tube

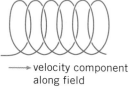

velocity component along field

Figure 6.28 Electrons moving in a helix

It is of interest to rotate the tube slightly, so that the velocity of the electrons is not normal to the magnetic field. In this case, the path of the electrons is seen to be a helix (rather like the coils of a spring). The component of the velocity normal to the field gives rise to circular motion. However, there is also a component of velocity along the direction of the field. There is no force on the electrons resulting from this component of velocity. Consequently, the electrons execute circular motion *and* move in a direction normal to the plane of the circle. The circle is 'pulled out' into a helix (see Figure 6.28). The helical path is an important aspect of the focusing of electron beams by magnetic fields in an electron microscope (Figure 6.29).

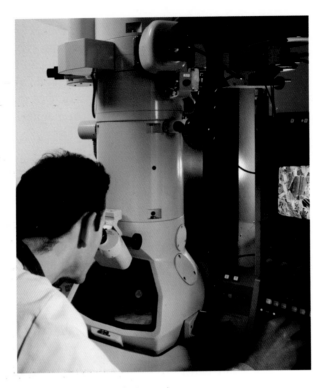

Figure 6.29 Laboratory technician using an electron microscope

Example

Electrons are accelerated to a speed of $8.4 \times 10^6 \, \text{m s}^{-1}$. They then pass into a region of uniform magnetic flux of flux density $0.50 \, \text{mT}$. The path of the electrons in the field is a circle with a radius $9.6 \, \text{cm}$. Calculate:

(a) the specific charge of the electron,

(b) the mass of the electron, assuming the charge on the electron is $1.6 \times 10^{-19} \, \text{C}$.

(a) $e/m = v/Br$
$= 8.4 \times 10^6/(0.50 \times 10^{-3} \times 0.096)$
$= \mathbf{1.8 \times 10^{11} \, C \, kg^{-1}}$

(b) $e/m = 1.8 \times 10^{11} = 1.6 \times 10^{-19}/m$
$m = \mathbf{9.1 \times 10^{-31} \, kg}$

Now it's your turn

Electrons are accelerated through a potential difference of $220 \, \text{V}$. They then pass into a region of uniform magnetic flux of flux density $0.54 \, \text{mT}$. The path of the electrons is normal to the magnetic field. Given that the charge on the electron is $1.6 \times 10^{-19} \, \text{C}$ and its mass is $9.1 \times 10^{-31} \, \text{kg}$, calculate:

(a) the speed of the accelerated electron,

(b) the radius of the circular path in the magnetic field.

Velocity selection of charged particles

We have seen that when particles of mass m and charge $+q$ enter an electric field of field strength E, there is a force F_E on the particle given by

$$F_E = qE$$

If the velocity of the particles before entry into the field is v at right angles to the field lines (see Figure 6.30), the particles will follow a parabolic path as they pass through the field.

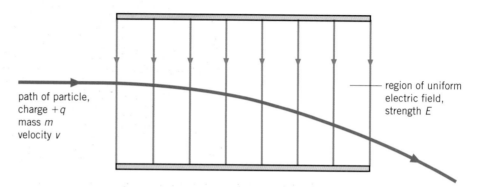

path of particle, charge $+q$ mass m velocity v

region of uniform electric field, strength E

Figure 6.30

Now suppose that a uniform magnetic field acts in the same region as the electric field. If this field acts downwards into the plane of the page, then, by Fleming's left-hand rule, a force will act on the charged particle in the direction opposite to the force due to the electric field. The magnitude F_B of this force is given by

$$F_B = Bqv$$

where B is the flux density of the magnetic field.

If the magnitude of one of the two fields is adjusted, then a situation can arise where the two forces, F_E and F_B, are equal in magnitude but opposite in direction. Thus

$$Bqv = qE$$

and

$$v = E/B$$

For the value of the velocity given by E/B, the particles will not be deflected, as shown in Figure 6.31. Particles with any other velocities will be deflected. If a parallel beam of particles enters the field then all the particles passing undeviated through the slit will have the same velocity. Note that the mass does not come into the equation for the speed and so, particles with a different mass (but the same charge) will all pass undeviated through the region of the fields if they satisfy the condition $v = E/B$.

The arrangement shown in Figure 6.31 is known as a velocity selector. Velocity selectors are very important in the study of ions. Frequently, the production of the ions gives rise to different speeds but to carry out investigations on the ions, the ions must have one speed only.

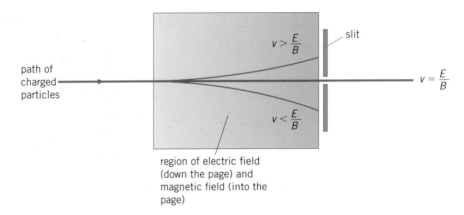

path of charged particles

$v > \dfrac{E}{B}$

slit

$v = \dfrac{E}{B}$

$v < \dfrac{E}{B}$

region of electric field (down the page) and magnetic field (into the page)

Figure 6.31 Velocity selector

Example

It is required to select charged ions which have a speed of $4.2 \times 10^6\,\text{m s}^{-1}$. The electric field strength in the velocity selector is $3.2 \times 10^4\,\text{V m}^{-1}$. Calculate the magnetic field strength required.

$v = E/B$
$B = 3.2 \times 10^4 / 4.2 \times 10^6$
$\quad = 7.6 \times 10^{-3}\,\text{T}$

Now it's your turn

Singly charged ions pass undeviated through a velocity selector. The electric field strength in the selector is $3.6 \times 10^4\,\text{V m}^{-1}$ and the magnetic flux density is 8.5 mT. Calculate the selected speed of the ions.

The mass spectrometer

An important application of the deflection of charged particles in magnetic and electric fields comes in the mass spectrometer. The mass spectrometer is a device used to measure the masses of atoms.

Ions of the atom under investigation are produced in the source S (see Figure 6.32) by heating a suitable chemical compound of that atom, or by passing an electric discharge through the vapour of the compound. These positive ions are attracted towards a slit S_1 (which must be at a negative potential if it is to attract the positive ions) so that a beam of ions of different speeds, and possibly different masses if the source has produced ions of different chemical elements, enters a velocity selector such as that described on page 121. If the electric field in the selector is E and the magnetic field is B_1, the velocity selected will be $v = E/B_1$. Note that, although ions of this particular speed v pass straight through the selector, the process is still an example of the deflection of charged particles by electric and magnetic fields. Ions of other speeds are deflected out of the straight-line path.

When this beam leaves the velocity selector part of the apparatus, it comes into a region where only a magnetic field B_2 acts. Here, the straight beam is

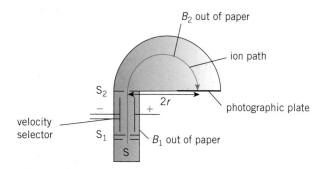

Figure 6.32 The mass spectrometer

deflected into a circular path. The relation between the mass m of the ion and the radius r of this circle is

$$m = (qB_1B_2r)/E$$

In some simplifications of this derivation you will find that it is assumed that the same magnetic field is applied in the velocity selector and magnetic deflection parts of the apparatus, although this is unlikely in practice. In these simple cases, the B_1B_2 term in the expression is replaced by B^2.

The ions are detected using a photographic plate or a very sensitive current detector. The position of the detector at which the maximum signal is obtained allows the measurement of the diameter, and hence the radius r of the circular path. In either case, everything on the right-hand side of the equation for m can be measured. This enables us to find the mass of the ion. Note that the radius of the path is proportional to the mass of the ion, for ions of the same charge.

Use of the mass spectrometer in early determinations of ionic mass led to an important discovery about the structure of nuclei. Even when it was certain that the ions produced from the source were those of a single element, it was sometimes found that the detector gave signals corresponding to circular paths of more than one radius. This could only happen if there were ions of the same element, but of different masses. These ions of the same element which differ in mass are called **isotopes**. We shall see what this means in terms of nuclear structure in Chapter 7. By comparing the intensities of the signals caused by different isotopes, the relative abundance of each isotope in the atoms in a given sample can be determined.

Mass spectrometers are also used to separate different isotopes for use in particular nuclear reactions.

Example

The most abundant isotope of carbon is carbon-12 (that is, the nuclei have mass 12.0 u). In a particular mass spectrometer experiment using carbon ions, it is found that a very strong signal is obtained from carbon ions that travel along a path of radius 25.1 cm. A much fainter signal is observed from ions that have followed a path of radius 29.3 cm. What can be deduced from this result?

The strong signal caused by ions in the 25.1 cm radius path must be due to ions of the most abundant isotope, carbon-12. Remember that the radius of the path is proportional to the mass of the ion; the 29.3 cm path must be followed by ions of mass m, where

$$m/12.0 = 29.3/25.1$$

Solving this proportionality relation gives $m = 14.0\,u$. Don't jump to the conclusion that this trace must be caused by ions of the carbon-14 isotope (mass 14.0 u); it may be, but unless you are sure that the source is producing carbon ions only, the trace could equally well be caused by nitrogen-14, the most abundant isotope of nitrogen.

Now it's your turn

In an early form of mass spectrometer ions of charge $+q$ were accelerated from rest through a potential difference $-V$. It was assumed that the ions were then all travelling with the same speed. The ions then entered a region of uniform magnetic field B, and the radius r of the path they followed was determined. Find an expression for the mass m of an ion, in terms of B, q, r and V. Suggest why later designs of mass spectrometer incorporated a velocity selector.

The Hall effect

Consider a thin slice of a conductor which is normal to a magnetic field, as illustrated in Figure 6.33. When there is a current in the conductor in the direction shown, charge carriers (electrons in a metal) will be moving at right angles to the magnetic field. They will experience a force which will tend to make them move to one side of the conductor. A potential difference known as the Hall voltage V_H will develop across the conductor. The Hall voltage does not increase indefinitely but reaches a constant value when the force on the charge carrier due to the magnetic field is equal to the force due to the electric field set up as a result of the Hall voltage.

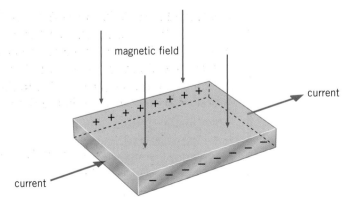

Figure 6.33 The Hall effect

The size of the Hall voltage depends on the material of the conductor, the current in the sample and the magnetic flux density. If the current is kept constant, then the Hall voltage across a sample will be proportional to the magnetic flux density. The Hall effect provides a means by which flux densities may be measured, using a Hall probe. In general, the material used for a Hall probe is a semiconductor. This gives a Hall voltage many thousands of times greater than that for a metal under the same conditions.

Comparing the effects of fields

We can summarise the effects of the different sorts of fields on masses, charges and current-carrying conductors. Although, as we have seen, there are close analogies between gravitational and electric fields, there are some ways in which they behave very differently.

Start with the effect of a gravitational field on a mass. Because masses always attract each other, a mass placed in a gravitational field will always move in the direction of the field, from a position of higher potential to a lower potential. For a field produced by a point mass, the field strength obeys an inverse square law relationship, and the potential obeys a reciprocal relationship with distance from the source of the field.

Electric fields are like gravitational fields in that, for a field produced by a point charge, the field strength is also given by the inverse square law, and the potential by a reciprocal relationship. However, we can have both positive charges and negative charges. A positive electric charge moves in the direction of the field, from a position of higher potential to a lower potential (just like a mass in a gravitational field). But a negative charge does just the opposite, against the direction of the field and from a low potential to a high potential.

What about electric charges in a magnetic field? A stationary charge is unaffected, but a moving charge experiences a force F given by $F = Bqv \sin \theta$. The direction of the force is given by Fleming's left-hand rule (for positive charges).

Finally, a current-carrying conductor in a magnetic field does not experience a force if the conductor is parallel to the field direction, but for all other directions it experiences a force given by $F = BIl \sin \theta$. The direction of the force is again given by Fleming's left-hand rule.

Example

A charged particle has mass 6.7×10^{-27} kg and charge $+3.2 \times 10^{-19}$ C. It is travelling at speed 2.5×10^8 m s^{-1} when it enters a region of space where there is a uniform magnetic field of flux density 1.6 T at right angles to its direction of motion. Calculate:
(a) the force on the particle due to the magnetic field,
(b) the radius of its orbit in the field.

(a) Force $= Bqv \sin \theta$
$$= 1.6 \times 3.2 \times 10^{-19} \times 2.5 \times 10^8 \times 1$$
$$= \mathbf{1.3 \times 10^{-10} \, N}$$

(b) Centripetal force is provided by the electromagnetic force

$$mv^2/r = Bqv$$
$$6.7 \times 10^{-27} \times (2.5 \times 10^8)^2/r = 1.3 \times 10^{-13}$$
$$\boldsymbol{r = 3.2\,m}$$

Now it's your turn

1 An electron of mass 9.1×10^{-31} kg and charge -1.6×10^{-19} C is travelling at a speed of 7.0×10^7 m s^{-1} when it enters a region of space where there is a uniform magnetic field of flux density 4.5 mT at right angles to its direction of motion. Calculate:
   (a) the force on the electron due to the magnetic field,
   (b) the radius of its orbit in the field.

2 An electron and an α-particle travelling at the same speed both enter the same region of uniform magnetic flux which is at right angles to their direction of motion. State and explain any differences between the two paths in the field.

Section 6.2 Summary

- There is a force on a current-carrying conductor whenever it is at an angle to a magnetic field.
- The direction of the force is given by Fleming's left-hand rule – place the first two fingers and thumb of the left hand at right-angles to each other, first finger in the direction of the magnetic field, second finger in the direction of the current, then the thumb gives the direction of the force.
- Magnetic flux density (field strength) is measured in tesla (T). 1 T = 1 Wb m^{-2}
- The magnitude of the force F on a conductor of length L carrying a current I at an angle θ to a magnetic field of flux density B is given by the expression: $F = BIl \sin \theta$
- The force between parallel current-carrying conductors provides a means by which the ampere may be defined in terms of SI units.
- The force F on a particle with charge q moving at speed v at an angle θ to a magnetic field of flux density B is given by the expression: $F = Bqv \sin \theta$
- The path of a charged particle, moving at constant speed in a plane at right-angles to a uniform magnetic field, is circular.

Section 6.2 Questions

1 A stiff straight wire has a mass per unit length of $45\,g\,m^{-1}$. The wire is laid on a horizontal bench and a student passes a current through it to try to make it lift off the bench. The horizontal component of the Earth's magnetic field is $18\,\mu T$ in a direction from south to north and the acceleration of free fall is $10\,m\,s^{-2}$.
   (a) (i) State the direction in which the wire should be laid on the bench.
       (ii) Calculate the minimum current required.
   (b) Suggest whether the student is likely to be successful with his experiment.

2 The magnetic flux density B at a distance r from a long straight wire carrying a current I is given by the expression

   $$B = 2.0 \times 10^{-7} \times \frac{I}{r}$$

   where r is in metres and I is in amps.
   (a) Calculate:
       (i) the magnetic flux density at a point distance $2.0\,cm$ from a wire carrying a current of $15\,A$,
       (ii) the force per unit length on a second wire, also carrying a current of $15\,A$, which is parallel to, and $2.0\,cm$ from, the first wire.
   (b) Suggest why the force between two wires is demonstrated in the laboratory using aluminium foil rather than copper wires.

3 A small horseshoe magnet is placed on a balance and a stiff wire is clamped in the space between its poles. The length of wire between the poles is $5.0\,cm$. When a current of $3.5\,A$ is passed through the wire, the reading on the balance increases by $0.027\,N$.
   (a) Calculate the magnetic flux density between the poles of the magnet.
   (b) State three assumptions which you have made in your calculation.

4 Figure 6.34 shows the track of a particle in a bubble chamber as it passes through a thin sheet of metal foil. A uniform magnetic field was applied into the plane of the paper.
   State with a reason:
   (a) in which direction the particle was moving,
   (b) whether the particle is positively or negatively charged.

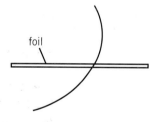

Figure 6.34

6.3 Electromagnetic induction

At the end of Section 6.3 you should be able to:
- define magnetic flux and the weber
- recall and solve problems using $\Phi = BA \cos \theta$
- define magnetic flux linkage

- infer from appropriate experiments on electromagnetic induction:
  - that a changing magnetic flux can induce an e.m.f. in a circuit
  - that the direction of the induced e.m.f. opposes the change producing it
  - the factors affecting the magnitude of the induced e.m.f.
- state and use Faraday's law of electromagnetic induction and Lenz's law
- explain simple applications of electromagnetic induction
- describe the function of a simple a.c. generator
- show an understanding of the principle of operation of a simple iron-cored transformer and recall and solve problems using $N_s/N_p = V_s/V_p = I_p/I_s$ for an ideal transformer
- show an appreciation of the scientific and economic advantages of alternating current and of high voltages for the transmission of electrical energy.

Magnetic flux

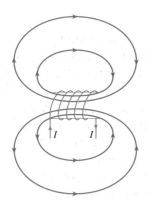

Figure 6.35 Continuous magnetic field lines

Figures 6.9, 6.11 and 6.12 illustrate the pattern of the magnetic field in the region of various conductors. However, the patterns are not all complete. All magnetic field lines should be continuous, as illustrated in Figure 6.35. Early experimenters thought that there was a flow of something along these lines and this gave rise to the idea of a magnetic flux, since 'flux' means 'flow'. Magnetic flux density (field strength) varies and this is shown by the spacing of the lines. Magnetic flux density may be considered as the number of lines of magnetic force per unit area of an area at right angles to the lines.

Figure 6.36 shows magnetic field lines passing through a coil of area A. The flux density of the field is B. The field lines approach the coil at an angle θ to a line normal to the plane of the coil. In the definition of magnetic flux, the component of the magnetic field perpendicular to the plane of the coil is $B \cos \theta$, so the magnetic flux density F is given by

$$\Phi = BA \cos \theta.$$

The magnetic flux Φ in a coil can be thought of as the total number of lines passing through the coil.

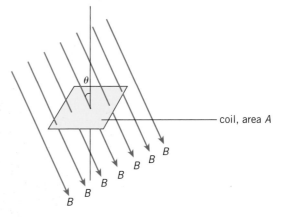

Figure 6.36 Magnetic field lines passing through a coil of area A at an angle θ to the normal to the plane of the coil. The magnetic flux density of the field is B. This gives a total flux Φ through the coil of $BA \cos \theta$.

Magnetic flux is the product of the magnetic flux density and the area normal to the lines of flux.

For a uniform magnetic field of flux density B which makes an angle θ with the normal to an area A, the magnetic flux Φ is given by the expression $\Phi = BA \cos \theta$.

The unit of magnetic flux is the **weber** (Wb). One weber is equal to one tesla metre-squared i.e. Tm^2.

The concept of magnetic flux is used frequently when studying electromagnetic induction.

Electromagnetic induction

The link between electric current and magnetic field was discovered by Oersted (1820). In 1831, Henry in the United States and Faraday in England demonstrated that an e.m.f. could be induced by a magnetic field. The effect was called **electromagnetic induction**.

Electromagnetic induction is now easy to demonstrate in the laboratory because sensitive meters are available. Figure 6.37 illustrates apparatus which may be used for this purpose. The galvanometer detects very small currents but it is important to realise that what is being detected are small electromotive forces (e.m.f.s). The current arises because there is a complete circuit incorporating an e.m.f. The following observations can be made.

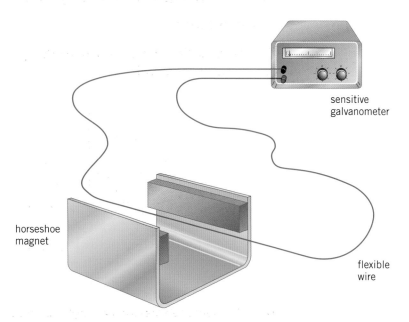

Figure 6.37 Apparatus to demonstrate electromagnetic induction

- An e.m.f. is induced when
  - the wire is moved through the magnetic field, across the face of the pole-pieces
  - the magnet is moved so that the wire passes across the face of the pole-pieces.
- An e.m.f. is *not* induced when
  - the wire is held stationary between the pole-pieces
  - the magnet is moved so that the pole-pieces move along the length of the wire
  - the wire moves lengthways so that it does not change its position between the poles of the magnet.

Figure 6.38 Wire wound to form a loop of several turns

These observations lead to the conclusion that an e.m.f. is induced whenever lines of magnetic flux are cut. The cutting may be caused by a movement of either the wire or of the magnet. The magnitude of the e.m.f. is also observed to vary.

- The magnitude of the e.m.f.
  - increases as the speed at which the wire is moved increases
  - increases as the speed at which the magnet is moved increases
  - increases, if the wire is made into a loop with several turns (see Figure 6.38)
  - increases as the number of turns on the loop increases.

It can be concluded that the magnitude of the induced e.m.f. depends on the rate at which magnetic flux lines are cut. The rate may be changed by varying the rate at which the flux lines are cut by a single wire or by using different numbers of turns of wire. The two factors are taken into account by using the term **magnetic flux linkage**. The change in magnetic flux linkage $\Delta(N\Phi)$ is equal to the product of the change in magnetic flux $\Delta\Phi$ and the number of turns N of a conductor involved in the change in flux.

change in magnetic flux linkage $\Delta(N\Phi) = N\Delta\Phi$

The experimental observations are summarised in **Faraday's law of electromagnetic induction**:

The e.m.f. induced is proportional to the rate of change of magnetic flux linkage.

The experimental observations made with the apparatus of Figures 6.37 and 6.38 have been concerned with the magnitude of the e.m.f. However, it is noticed that the direction of the induced e.m.f. changes and that the direction is dependent on the direction in which the magnetic flux lines are being cut. The direction of the induced e.m.f. or current in a wire moving through a magnetic field at right angles to the fields may be determined using **Fleming's right-hand rule**. This rule is illustrated in Figure 6.39.

If the first two fingers and thumb of the right hand are held at right angles to one another, the **F**irst finger in the direction of the magnetic **F**ield and the thu**M**b in the direction of **M**otion, then the se**C**ond finger gives the direction of the induced e.m.f. or **C**urrent.

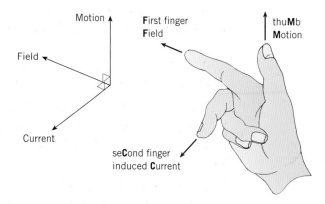

Figure 6.39 Fleming's right-hand rule

An explanation for the direction of the induced e.m.f. can be found by reference to the motor effect and conservation of energy.

Figure 6.40 shows a wire being moved downwards through a magnetic field. Since the wire is in the form of a continuous loop, the induced e.m.f. gives rise to a current, and the direction of this current can be found using Fleming's right-hand rule. This current is at right-angles to the magnetic flux and, by the motor effect, there will be a force on the wire. Using Fleming's left-hand rule, the force is upwards when the wire is moving downwards. Reversing the direction of motion of the wire causes a current in the opposite direction and hence the electromagnetic force would once again oppose the motion. This conclusion is not surprising when conservation of energy is considered. An electric current is a form of energy and this energy must have been converted from some other form. Movement of the wire against the electromagnetic force means that work has been done in overcoming this force and it is this work which is seen as electrical energy. Anyone who has ridden a bicycle with a dynamo will realise that work has to be done to light the lamp!

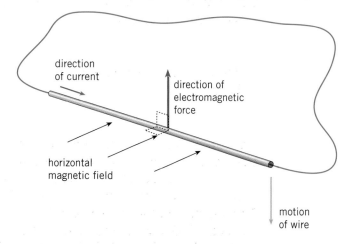

Figure 6.40

This application of conservation of energy is summarised in **Lenz's law**:

> The direction of the induced e.m.f. is such as to cause effects to oppose the change producing it.

Faraday's law of electromagnetic induction and Lenz's law may be summarised using the equation

$$E = -\frac{d(N\Phi)}{dt}$$

(see the Maths Note below) where E is the e.m.f. induced by a rate of change of flux linkage of $d(N\Phi)/dt$. The minus sign indicates that the induced e.m.f. causes effects to oppose the change producing it.

Maths Note

The shorthand way of expressing the rate of change of a quantity x with time t is

$$\text{rate of change of } x \text{ with } t = \frac{dx}{dt}$$

This represents a mathematical operation known as *differentiation*. It is achieved by finding the gradient of the graph of x against t.

You will come across this notation here, in connection with the rate of change of magnetic flux linkage, and in one of the equations for radioactive decay (see Chapter 7).

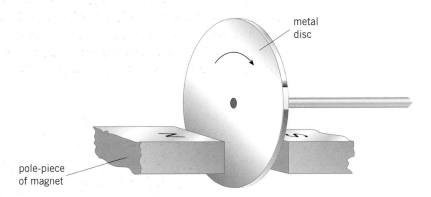

Figure 6.41 Apparatus to demonstrate eddy current damping

The conversion of mechanical energy to electrical energy may be shown by spinning a metal disc in a magnetic field, as illustrated in Figure 6.41. The disc is seen to slow down much more rapidly with the magnet in place than when it has been removed. As the disc spins, it cuts through the flux lines of the magnet. This cutting becomes more obvious if a radius of the disc is considered. As the radius rotates, it will cut flux. An e.m.f. will be induced in the disc but, because the rate of cutting of flux varies from one part of the disc to another, the e.m.f. will have different magnitudes in different regions. The disc is metal and therefore electrons will move between regions with different e.m.f. values. Currents are induced in the disc. Since these currents vary in magnitude and direction there are called **eddy currents**. The eddy currents cause heating in the disc and the dissipation of the energy of rotation of the disc is referred to as **eddy current damping**. If the permanent magnet in Figure 6.41 is replaced by an electromagnet, the spinning disc will be slowed down whenever there is a current in the electromagnet. This is the principle behind electromagnetic braking. The advantage over conventional brakes is that there is no physical contact with the spinning disc. This makes such brakes very useful for trains travelling at high speeds. However, the disadvantage is that, as the disc slows down, the induced eddy currents will be smaller and therefore the braking will be less efficient. This system would be useless as the parking brake on a car!

The a.c. generator

The application of electromagnetic induction that has had most effect on our everyday lives is the a.c. generator. Imagine life without electricity being available at the touch of a switch! An electric generator converts mechanical energy, in the form of the rotational energy of a coil of wire, into electrical energy.

Figure 6.42 is a sketch of the important parts of an a.c. generator. A rectangular coil of wire is mounted on an axle so that the coil can rotate in a magnetic field. The axle is at right angles to the direction of the field. The axle is turned by some mechanical means, such as a steam turbine, so that it has rotational energy. The sides of the coil that are parallel to the axle (and hence at right angles to the magnetic field) cut the magnetic field when the axle rotates, so that an e.m.f. is induced in these sides. The other sides of the coil do not cut the field, and there is no induced e.m.f. in these sides. The polarities of the e.m.f.s add up going round the coil, so that the resultant effect is that a current is generated in an external load. This current is taken from the coil and fed to the external circuit by brushes pressing against slip-rings on the axle. If you follow the polarities of the e.m.f.s in the two wires

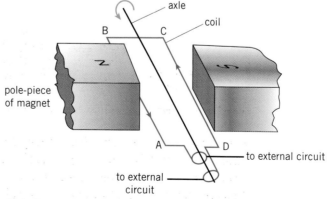

Figure 6.42 The a.c. generator
When the axle rotates, an e.m.f. is induced in the sides AB and CD of the coil. No e.m.f. is induced in sides BC and DA.

that cut the field, you will find that the direction of the current taken out through the slip-rings changes after every half-revolution. The generator thus produces an alternating e.m.f., or an alternating current if it is connected to a load circuit. The positions of zero e.m.f., where the e.m.f. is about to change direction, are when the plane of the coil is at right angles to the field. In these positions, the wires are not cutting the field lines, so that no e.m.f. is induced. For the positions of maximum induced e.m.f., the plane of the coil is in the direction of the field; that is every half-revolution.

You are not expected to derive the formula for the induced e.m.f., but it is interesting to see its form. For a coil of N turns and area A rotating at an angular speed ω in a uniform magnetic field of flux density B, the induced e.m.f. E is given by

$$E = NAB\omega \sin \omega t = E_0 \sin \omega t$$

where E_0 is the peak value of the output e.m.f. of the generator. This equation shows that the output is sinusoidal, of angular frequency ω equal to the angular frequency of rotation of the coil. The frequency of an alternating e.m.f. is usually measured in hertz rather than in radians per second, so the frequency f of the output is $\omega/2\pi$. The equation agrees with our prediction that the output is sinusoidal, as shown in Figure 6.43.

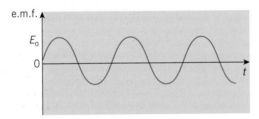

Figure 6.43 a.c. generator output

It would also be expected that the induced e.m.f. (which is proportional to the rate of change of magnetic flux linkage) would be proportional to the magnitude of the magnetic flux density of the field in which the coil is rotating and to the rate of rotation of the coil. It is also no surprise that the maximum induced e.m.f. is proportional to the number of turns in the coil and to the area of the coil, as the magnetic flux linkage depends on the product of these factors as well as on the magnetic flux density.

Example

The uniform flux density between the poles of a magnet is 0.080 T. A small coil of area of cross-section 6.5 cm^2 has 250 turns and is placed with its plane normal to the magnetic field. The coil is withdrawn from the field in a time of 0.26 s.

Determine:
(a) the magnetic flux through the coil when it is between the poles of the magnet,

(b) the change in magnetic flux linkage when the coil is removed from the field,

(c) the average e.m.f. induced in the coil whilst it is being withdrawn.

(a) magnetic flux $\Phi = BA \sin \theta$
$$= 0.080 \times 6.5 \times 10^{-4}$$
$$= \mathbf{5.2 \times 10^{-5} \, Wb}$$

(b) change in flux linkage $= (N\Phi)_{\text{FINAL}} - (N\Phi)_{\text{INITIAL}}$
$$= 0 - (250 \times 5.2 \times 10^{-5})$$
$$= \mathbf{-1.3 \times 10^{-2} \, Wb}$$

(the sign indicates that the flux linkage is decreasing)

(c) induced e.m.f. $= \dfrac{\text{change in flux linkage}}{\text{time taken}}$

$$= (1.3 \times 10^{-2})/0.26$$
$$= \mathbf{0.050 \, V}$$

Now it's your turn

1 An aircraft has a wingspan of 17 m and is flying horizontally in a northerly direction at a speed of 94 m s^{-1}. The vertical component of the Earth's magnetic field is 40 μT in a downward direction.
(a) Calculate:
    (i) the area swept out per second by the wings,
    (ii) the magnetic flux cut per second by the wings,
    (iii) the e.m.f. induced between the wingtips.
(b) State which wing-tip will be at the higher potential.

2 A current-carrying solenoid produces a uniform magnetic flux of density 4.6×10^{-2} T along its axis. A small circular coil of radius 1.2 cm has 350 turns of wire and is placed on the axis of the solenoid with its plane normal to the axis. Calculate the average e.m.f. induced in the coil when the current in the solenoid is switched off in a time of 85 ms.

3 A metal disc is made to spin at 15 revolutions per second about an axis through its centre normal to the plane of the disc. The disc has radius 24 cm and spins in a uniform magnetic field of flux density 0.15 T, parallel to the axis of rotation.
Calculate:
    (i) the area swept out each second by a radius of the disc,
    (ii) the flux swept out each second by a radius of the disc,
    (iii) the e.m.f. induced in the disc.

e.m.f. induced between two coils

A current-carrying solenoid or coil is known to have a magnetic field. If the current in the coil is switched off, there will be a change in flux in the region around the coil. Consider the apparatus illustrated in Figure 6.44.

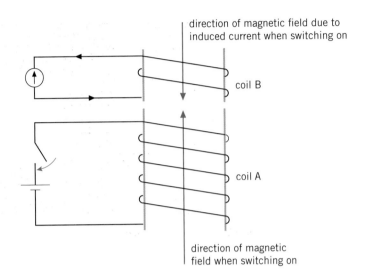

direction of magnetic field due to induced current when switching on

coil B

coil A

direction of magnetic field when switching on

Figure 6.44

As the current in coil A is being switched on, the magnetic field in this coil grows. The magnetic field links with the turns on coil B and, as a result, there is a change in flux linkage in coil B and an e.m.f. is induced in this coil. Coil B forms part of a complete circuit and hence there is a current in coil B. The direction of this current can be determined using Lenz's law.

The change which brought about the induction of a current was a *growth* in the magnetic flux in coil A. The induced current in coil B will give rise to a magnetic field in coil B and this field will, by Lenz's law, try to oppose the growth of the field in coil A. Consequently, since the field in coil A is vertically upwards (the right-hand grip rule), the field in coil B will be vertically downwards and the induced current will be in an anticlockwise direction through the meter. When the current in coil A is switched off, the magnetic field in coil A will *decay*. The magnetic field in coil B due to the induced current must try to prevent this decay and hence it will be vertically upwards. The induced current has changed in direction.

The magnitude of the induced e.m.f. can be increased by inserting a soft iron core into the coil (but be careful not to damage the meter as any induced e.m.f. will be very much greater) or by increasing the number of turns on the coils or by switching a larger current in coil A.

It is important to realise that an e.m.f. is induced only when the magnetic flux in coil A is changing; that is, when the current in coil A is changing. A steady current in coil A will not give rise to an induced e.m.f. An e.m.f. may be induced continuously in coil B if an alternating current is provided for coil A. This is the principle of the **transformer**.

A simple transformer is illustrated in Figure 6.45. Two coils of insulated wire are wound on to a laminated soft-iron core. An alternating e.m.f. is applied across one coil (the primary coil) and an e.m.f. is induced in the secondary coil. The ratio of the output e.m.f. to the applied e.m.f. is dependent on the ratio of the number of turns on the two coils. The applied e.m.f. and the induced e.m.f. across the secondary coil have the same frequency but are out of phase by π rad (180°).

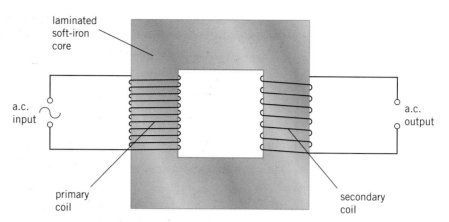

Figure 6.45 Simple transformer

Consider a transformer with N_p turns and N_s turns on the primary and secondary coils respectively, with currents in the coils of I_p and I_s when the e.m.f.s across the coils and V_p and V_s.

For a transformer which is 100% efficient (an **ideal** transformer),

input power = output power
$$V_p I_p = V_s I_s$$

and

$$\frac{N_s}{N_p} = \frac{V_s}{V_p} = \frac{I_p}{I_s}$$

When V_s is greater than V_p, there are more turns on the secondary coil than the primary and the transformer is said to be a **step-up transformer**. Conversely, if V_s is less than V_p, the secondary coil has fewer turns than the primary and the transformer is said to be a **step-down transformer**.

In practice, the transformer will not be 100% efficient due to power losses. Some sources of these losses are:

■ loss of magnetic flux between the primary and secondary coils
■ resistive heating in the primary and secondary coils
■ heating of the core due to eddy currents
■ heating of the core due to repeated magnetisation and demagnetisation.

The use of soft iron reduces heating due to magnetisation and demagnetisation of the core. However, since iron (an electrical conductor) is used as the material for the core so that the magnetic flux linkage is large, eddy currents cannot be prevented. Laminating the core, i.e. building up the core from thin strips of soft iron which are electrically insulated from one another, reduces eddy current losses.

Figure 6.46 Transformers are produced in many shapes and sizes

Example

A transformer is to be used with an alternating supply of 240 V to power a lamp rated as 12 V, 3 A. The secondary coil of the transformer has 80 turns. Assuming that the transformer is ideal, calculate, for the primary coil:

(a) the number of turns,
(b) the current.

(a) For an ideal transformer,

$$V_s/V_p = N_s/N_p$$
$$12/240 = 80/N_p$$
$$N_p = \textbf{1600 turns}$$

(b) For an ideal transformer,

$$V_s/V_p = I_p/I_s$$
$$12/240 = I_p/3$$
$$I_p = \textbf{0.15 A}$$

Now it's your turn

1 The primary coil of an ideal transformer has 1200 turns and is connected to a 240 V alternating supply. The transformer is to be used to step down the voltage to 9.0 V.

  (a) Calculate the number of turns on the secondary coil.
  (b) An appliance, rated as 9.0 V, 1.5 A is connected to the secondary coil of the transformer. Calculate the current in the primary coil.

2 An ideal transformer is to be used to step up a voltage from 220 V to supply a current of 15 mA at 6.6 kV. The primary coil has 250 turns. Calculate:

  (a) the ratio of the number of turns on the secondary coil to that on the primary coil,
  (b) the number of turns on the secondary coil,
  (c) the current in the primary coil.

Transmission of electrical energy

Transformers play an essential part in the transmission of electrical energy. Generating stations are often situated a long way from the cities they serve. Hydroelectric plants must be placed at the dam providing the water supply, and nuclear generators need a plentiful supply of cooling water; there is also a safety factor to consider. Fossil-fuel plants are also placed some distance from towns, mainly because of air pollution concerns and because land is cheaper away from towns. This means that, whatever the means of generating the electricity, it is likely to need to be transmitted over long distances. There is always a power loss in transmission lines, simply due to the I^2R effect (see Chapter 5 of the AS book). However, it turns out that this loss can be reduced if the power is transmitted at a high voltage. Look at the following Example.

Example

A small town, several km from an electricity generating station, requires 120 kW of power on average. The total resistance of the transmission lines is 0.40 Ω. Calculate the power loss if the transmission is made at a voltage of **(a)** 240 V, **(b)** 24 kV.

First calculate the current I in the transmission lines in each case. In (a), $I = P/V = 1.2 \times 10^5/240 = 500$ A. In (b), the same calculation gives $I = 5.0$ A.

Now find the power loss from $P = I^2R$. In (a), it is $(500)^2 \times 0.40 = $ **100 kW**, or more than 80% of the power required by the town. In (b), the power loss is $5.0^2 \times 0.40 = $ **10 W**, less than one-hundredth of 1% of the power requirement.

The advantage of transmitting power at the higher voltage is now obvious. The advantage of using alternating current rather than direct current is that a.c. is so efficiently stepped up to a high voltage for transmission using a transformer. Transformers used in transmission networks have very high efficiencies, often over 98%.

Section 6.3 Summary

- Magnetic field is a general term used for a region of space where a current-carrying conductor experiences an electromagnetic force.
- Magnetic flux density (magnetic field strength) measures the strength of a magnetic field and has a magnitude and direction at every point in the field.
- Magnetic flux is the product of flux density and area normal to the flux.
- The direction of the induced current in a conductor moving through a magnetic field is given by Fleming's right-hand rule. That is:
  If the first two fingers and thumb of the right hand are held at right-angles to each other, the first finger in the direction of the magnetic field and the thumb in the direction of motion, then the second finger gives the direction of the induced e.m.f. or current.
- Faraday's law of electromagnetic induction states that the e.m.f. induced is proportional to the rate of change of magnetic flux linkage.
- Lenz's law states that the direction of the induced e.m.f. is such as to cause effects to oppose the change producing it.
- Faraday's law of electromagnetic induction and Lenz's law may be summarised using the equation:

  $E = -\mathrm{d}(N\Phi)/\mathrm{d}t$

  where E is the e.m.f. induced by a rate of change of flux linkage of $\mathrm{d}(N\Phi)/\mathrm{d}t$. The sign indicates the relative direction of the e.m.f. and the change in flux linkage.

In a transformer, an alternating current in one coil induces an alternating e.m.f. in a second coil. The ratio of the output (secondary) e.m.f. to the applied (primary) e.m.f. equals the ratio of the number of turns on the secondary coil to the number of turns on the primary coil:

$$V_s/V_p = N_s/N_p$$

Section 6.3 Questions

1 A coil is constructed by winding 400 turns of wire on to a cylindrical iron core. The mean radius of the coil is 3.0 cm. It is found that the flux density B in the core due to a current I in the coil is given by the expression

$B = 2.2\ I$

where B is in tesla.
(a) Calculate, for a current of 0.64 A in the coil:
   (i)   the magnetic flux density in the core,
   (ii) the magnetic flux in the core,
   (iii) the flux linkage of the coil.
(b) The current in the coil is switched off in a time of 0.011 s. Calculate the e.m.f. induced and state where this e.m.f. will be observed.
(c) Hence suggest why, when switching off a large electromagnet, the current is reduced gradually rather than switched off suddenly.

2 A flat coil contains 250 turns of insulated wire and has a mean radius of 1.5 cm. The coil is placed in a region of uniform magnetic flux of flux density 85 mT such that there is an angle θ between the plane of the coil and the flux lines. The coil is withdrawn from the magnetic field in a time of 0.30 s.
Calculate the average e.m.f. induced in the coil for an angle θ of:
(a) zero,
(b) 90°,
(c) 35°.

3 An ideal transformer is to be used to step down a 220 V alternating supply to 7.5 V.

(a) Calculate the ratio of the number of turns on the secondary coil to that on the primary coil.
(b) Calculate the number of turns on the primary coil given that there are 140 turns on the secondary coil.
(c) Suggest why the wire forming the secondary coil would, in this case, be thicker than that for the primary coil.

Exam-style Questions

1 A long straight current-carrying wire is held in a north-south direction vertically above a small compass. The vertical distance between the compass and the wire is 4.6 cm.

   The horizontal component of the Earth's field is 18 μT and the magnetic flux density B due to a current I in a long straight wire is given by the expression

$$B = 2.0 \times 10^{-7} \times \frac{I}{r}$$

   where B is measured in tesla and r is the distance from the wire in metres.
   (a) State the direction in which the compass points when there is no current in the wire.
   (b) A steady current of 6.0 A is switched on in the wire.
      (i) State the direction of the magnetic field of the wire in the region of the compass.
      (ii) Calculate the magnetic flux density due to the current in the wire at the position of the compass.
      (iii) Determine the angle through which the compass will rotate when the current in the wire is switched on.
   (c) The compass is replaced by a long straight wire so that this wire is parallel to and 4.5 cm from the first wire. This second wire carries a current of 7.7 A. Calculate the force per unit length between the two wires.

2 The magnetic flux density B at the centre of a long solenoid carrying a current I is given by the expression

$$B = 2.5 \, I$$

   where B is in millitesla and I is in amps. The cross-section of the solenoid has a diameter of 2.5 cm. A length of wire is wound tightly round the centre of the solenoid to form a small coil of 50 turns, as illustrated in Figure 6.47.
   (a) Calculate, for a current of 1.8 A in the solenoid:
      (i) the flux density in the solenoid,
      (ii) the magnetic flux linkage of the small coil.

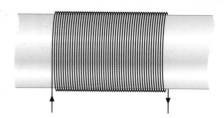

Figure 6.47

(b) The current is reversed in direction in 0.031 s. Calculate the average e.m.f. induced in the small coil.

3 Two coils A and B are placed as shown in Figure 6.48.

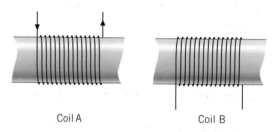

Coil A   Coil B

Figure 6.48

The current in coil A is switched on and then off so that the variation with time t of the current I is as shown in Figure 6.49.

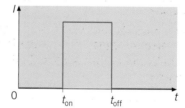

Figure 6.49

Sketch a graph to show the variation with time t of the e.m.f. E induced in coil B.

7. Nuclear physics

In the early 1900s, Rutherford showed that the atom consists of a very small nucleus containing the majority of the atom's mass, surrounded by orbiting electrons. In this chapter, we shall look at some of the physics of the nucleus. We start by looking at the structure of the nucleus. The decay of unstable nuclei leads to radioactive emissions. We shall investigate the properties of these emissions. Nuclei may also be transformed by particle impact. We shall look at the processes of fission and fusion, with particular reference to energy changes.

7.1 Atomic structure

At the end of Section 7.1 you should be able to:

- describe a simple model for the nuclear atom to include protons, neutrons and orbital electrons
- distinguish between nucleon number (mass number) and proton number (atomic number)
- show an understanding that an element can exist in various isotopic forms, each with a different number of neutrons
- use the usual notation for the representation of nuclides.

The atoms of all elements are made up of three particles called **protons**, **neutrons**, and **electrons**. The protons and neutrons are at the centre or nucleus of the atom. The electrons orbit the nucleus.

We shall see later that the diameter of the nucleus is only about 1/10 000 of the diameter of an atom.

Figure 7.1 illustrates very simple models of a helium atom and a lithium atom.

The protons and neutrons both have a mass of about one atomic mass unit (1 u = 1.66×10^{-27} kg). By comparison, the mass of an electron is very small, about 1/2000 of 1 u. The vast majority of the mass of the atom is therefore in the nucleus.

The basic properties of the proton, neutron and electron are summarised in Table 7.1.

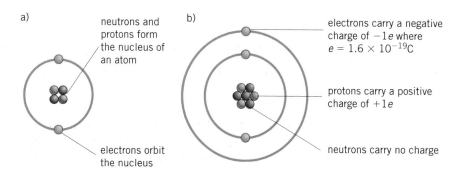

Figure 7.1 Structures of a) a helium atom and b) a lithium atom

Table 7.1

	approximate mass	charge	position
proton	u	$+e$	in nucleus
neutron	u	0	in nucleus
electron	u/2000	$-e$	orbiting nucleus

Atoms and ions

Atoms are uncharged because they contain equal numbers of protons and electrons. If an atom loses one or more electrons so that it does not contain an equal number of protons and electrons, it becomes charged and is called an **ion**.

For example, if a sodium atom loses one of its electrons it becomes a positive sodium ion.

$$\text{Na} \quad \rightarrow \quad \text{Na}^+ \quad + \quad \text{e}^-$$
sodium atom   sodium ion   electron

If an atom gains an electron, it becomes a negative ion.

Proton number and nucleon number

The number of protons in the nucleus of an atom is called the **proton number** (or **atomic number**) Z.

The number of protons together with the number of neutrons in the nucleus is called the **nucleon number** (or **mass number**) A.

A **nucleon** is the name given to either a proton or a neutron. The difference between the nucleon number (A) and the proton number (Z) gives the number of neutrons in the nucleus. Because atoms are electrically neutral, the proton number is equal to the number of electrons orbiting the nucleus.

Nuclear representation

The nucleus of one form of sodium contains 11 protons and 12 neutrons. Therefore its proton number Z is 11 and the nucleon number A is $11 + 12 = 23$. This nucleus can be shown as $^{23}_{11}$Na. In general, a nucleus can be represented as

$$^{\text{nucleon number}}_{\text{proton number}}X$$

where X is the chemical symbol.

> ## Example
>
> An oxygen nucleus is represented by $^{16}_{8}$O. Describe its atomic structure.
>
> The nucleus has a proton number of 8 and a nucleon number of 16. Thus, its nucleus contains **8 protons** and $16 - 8 = $ **8 neutrons**. There are also **8 electrons** (equal to the number of protons) orbiting the nucleus.
>
> ### Now it's your turn
>
> Write down the proton number and the nucleon number for the potassium nucleus $^{40}_{19}$K. Deduce the number of neutrons in the nucleus.

Isotopes

Sometimes atoms of the same element have different numbers of neutrons in their nuclei. The most abundant form of chlorine contains 17 protons and 18 neutrons in its nucleus, giving it a nucleon number of $17 + 18 = 35$. This is often called chlorine-35. Another form of chlorine contains 17 protons and 20 neutrons in the nucleus, giving it a nucleon number of 37. This is chlorine-37. Chlorine-35 and chlorine-37 are said to be **isotopes** of chlorine.

> Isotopes are different forms of the same element which have the same numbers of protons but different numbers of neutrons in their nuclei.

Some elements have many isotopes, but others have very few. For hydrogen, the most common isotope is hydrogen-1. Its nucleus is a single proton. Hydrogen-2 is called deuterium; its nucleus contains one proton and one neutron. Hydrogen-3, with one proton and two neutrons, is called tritium.

Note that the term isotope is also used to describe *nuclei* with the same proton number (that is, nuclei of the same element) but with different nucleon numbers. You may also come across the term **nuclide**.

> A nuclide is one type of nucleus with a particular nucleon number and a particular proton number.

Section 7.1 Summary

- An atom consists of a nucleus containing protons and neutrons surrounded by orbiting electrons.
- Most of the mass of an atom is contained in its nucleus.
- Neutral atoms contain equal numbers of protons and electrons.
- Atoms which have gained or lost electrons are charged, and are called ions.
- The nucleon number A of a nucleus is the number of nucleons (protons and neutrons) in the nucleus.
- The proton number Z of a nucleus is the number of protons in the nucleus; hence the number of neutrons in the nucleus is $A - Z$.
- A nucleus (chemical symbol X) may be represented by:

$$_{\text{proton number}}^{\text{nucleon number}}\text{X}$$

- Isotopes are different forms of the same element (that is, with the same proton number) but with a different nucleon number.

Sections 7.1–7.2 Questions

1 What do different isotopes of the same element have in common? How do they differ?

2 Distinguish between the terms *isotope* and *nuclide*.

3 Distinguish between the terms *nuclide* and *nucleus*.

7.2 α-particles, β-particles and γ-radiation

At the end of Sections 7.2–7.4 you should be able to:

- show an understanding of the nature of α-particles, β-particles and γ-radiation
- appreciate that nucleon number, proton number, and energy and mass are all conserved in nuclear processes
- represent simple nuclear reactions by nuclear equations of the form $_{7}^{14}\text{N} + _{2}^{4}\text{He} \rightarrow _{8}^{17}\text{O} + _{1}^{1}\text{H}$
- show an appreciation of the spontaneous and random nature of nuclear decay
- infer the random nature of radioactive decay from the fluctuations in count-rate.

Some elements have nuclei which are unstable. In order to become more stable, they emit particles and/or electromagnetic radiation. The nuclei are said to be **radioactive**, and the emission is called **radioactivity**. The emissions are invisible to the eye, but their tracks were first made visible in a device called a cloud chamber. The photograph shows tracks created by one type of emission, α-particles.

Examination of the nature and properties of the emitted particles or radiation shows that the emissions are of three different types. These are **α-particles** (alpha-particles), **β-particles** (beta-particles) and **γ-radiation** (gamma-radiation). All three emissions orginate from the nucleus.

Figure 7.2 Tracks of α-particles

α-particles

An α-particle is identical to the nucleus of a helium atom.

Like a helium atom, an α-particle contains two protons and two neutrons, and hence carries a charge of $+2e$. α-particles travel at speeds of up to about $10^7 \, \text{m s}^{-1}$ (about 5% of the speed of light). α-particle emission is the least penetrating of the three types of emission. It can pass through very thin paper, but is unable to penetrate thin card. Its range in air is a few centimetres. Because α-particles are charged, they can be deflected by electric and magnetic fields.

As α-particles travel through matter, they interact with nearby atoms causing them to lose one or more electrons. The ionised atom and the dislodged electron are called an ion pair. The production of an ion pair requires the separation of unlike charges, and this process requires energy. α-particles have a relatively large mass and charge, and consequently they are efficient ionisers. They may produce as many as 10^5 ion pairs for every centimetre of air through which they travel. Thus they lose energy relatively quickly, and have low penetrating power.

When the nucleus of an atom emits an α-particle, it is said to undergo α-decay. The nucleus loses two protons and two neutrons in this emission.

In α-decay the proton number of the nucleus decreases by two, and the nucleon number decreases by four.

Each element has a particular proton number, and therefore α-decay causes one element to change into another. (This process is sometimes called **transmutation**.) The original nuclide is called the **parent** nuclide, and the new one the **daughter** nuclide.

For example, uranium-234 (the parent nuclide) may emit an α-particle. The daughter nuclide is thorium-230. In addition, energy is released. This emission is represented by the nuclear equation

$$^{234}_{92}\text{U} \rightarrow {}^{230}_{90}\text{Th} + {}^{4}_{2}\text{He} + \text{energy}$$

β-particles

β-particles are fast-moving electrons.

β-particles have speeds in excess of 99% of the speed of light. These particles have half the charge and very much less mass than α-particles. Consequently, they are much less efficient than α-particles in producing ion pairs. They are thus far more penetrating than α-particles, being able to travel up to about a metre in air. They can penetrate card and sheets of aluminium up to a few millimetres thick. Their charge means that they are affected by electric and magnetic fields. However, there are important differences between the behaviour of α- and β-particles in these fields. β-particles carry a negative charge, and are thus deflected in the opposite direction to the positively charged α-particles. Because the mass of a β-particle is much less than that of an α-particle, β-particles experience a much larger deflection when moving at the same speed as α-particles.

It was stated in Section 7.1 that the nucleus contains protons and neutrons. What, then, is the origin of β-particle emission? The β-particles certainly come from the nucleus, not from the electrons outside the nucleus. What happens is that just prior to β-emission a neutron in the nucleus forms a proton and an electron, in order to change the ratio of protons to neutrons in the nucleus. This makes the daughter nucleus more stable.

In fact, *free* neutrons are known to decay like this:

$$^{1}_{0}\text{n} \rightarrow {}^{1}_{1}\text{p} + {}^{0}_{-1}\text{e} + \text{energy}$$

A similar process happens in the nucleus. In β-decay the electron is emitted from the nucleus as a β-particle. This leaves the nucleus with the same number of nucleons as before, but with one extra proton and one fewer neutrons.

In β-decay a daughter nuclide is formed with the proton number increased by one, but with the same nucleon number.

For example, strontium-90 (the parent nuclide) may decay with the emission of a β-particle to form the daughter nuclide yttrium-90. As in the case of α-emission, energy is evolved. The decay is represented by the nuclear equation

$$^{90}_{38}\text{Sr} \rightarrow \ ^{90}_{39}\text{Y} + \ ^{0}_{-1}\text{e} + \text{energy}$$

γ-radiation

γ-radiation is part of the electromagnetic spectrum with wavelengths between 10^{-11} m and 10^{-13} m.

Since γ-radiation has no charge, its ionising power is much less than that of either α- or β-particles. γ-radiation penetrates almost unlimited thicknesses of air, several metres of concrete or several centimetres of lead.

α- and β-particles are emitted by unstable nuclei which have excess energy. The emission of these particles results in changes in the ratio of protons to neutrons, but the nuclei may still have excess energy. The nucleus may return to its unexcited (or ground) state by emitting energy in the form of γ-radiation.

In γ-emission no particles are emitted and there is therefore no change to the proton number or nucleon number of the parent nuclide.

For example, when uranium-238 decays by emitting an α-particle, the resulting nucleus of thorium-234 contains excess energy (it is in an excited state) and emits a photon of γ-radiation to return to the ground state. This process is represented by the nuclear equation

$$^{234}_{90}\text{Th}^* \rightarrow \ ^{234}_{90}\text{Th} + \ ^{0}_{0}\gamma$$

The * next to the symbol Th on the left-hand side of the equation shows that the thorium nucleus is in an excited state.

Note that in all radioactive decay equations (and, in fact, in all equations representing nuclear reactions) nucleon number and proton number are conserved. The sum of the numbers at the top of the symbols on the left-hand side of the equation (the sum of the nucleon numbers) is equal to the sum of the nucleon numbers on the right-hand side. Similarly, the sum of the numbers at the bottom of the symbols on the left-hand side (the sum of the proton numbers) is equal to the sum of the proton numbers on the right-hand side. Energy and mass, taken together, are also conserved in all nuclear processes.

Summary of the properties of radioactive emissions

Table 7.2 summarises the properties of α-particles, β-particles and γ-radiation.

Table 7.2

property	α-particle	β-particle	γ-radiation
mass	4 u	about u/2000	0
charge	$+2e$	$-e$	0
nature	helium nucleus (2 protons + 2 neutrons)	electron	short-wavelength electromagnetic waves
speed	up to $0.05c$	more than $0.99c$	c
penetrating power	few cm of air	few mm of aluminium	few cm of lead
relative ionising power	10^4	10^2	1
affects photographic film?	yes	yes	yes
deflected by electric, magnetic fields?	yes, see Figure 7.3	yes, see Figure 7.3	no

Figure 7.3 illustrates a hypothetical demonstration of the effect of a magnetic field on the three types of emission. The direction of the magnetic field is perpendicularly into the page. γ-radiation is uncharged, and is not deflected by the magnetic field. Because α- and β-particles have opposite charges, they are deflected in opposite directions. Note that the deviation of the α-particles is less than that of the β-particles, and that α-particles all deviate by the same amount. β-particles by contrast have a range of deflections indicating that they have a range of energies.

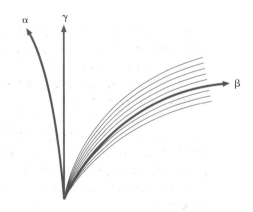

Figure 7.3 Deflection of α- and β-particles and γ-radiation by a magnetic field into the page

7.3 Radioactive decay series

The daughter nuclide of a radioactive decay may itself be unstable and so may emit radiation to give another different nuclide. This sequence of radioactive decay from parent nuclide through succeeding daughter nuclides is called a **radioactive decay series**. The series ends when a stable nuclide is reached.

Part of such a radioactive decay series, the uranium series, is shown in Table 7.3.

Table 7.3 Part of the decay series of uranium-238

decay	radiation emitted
$^{238}_{92}\text{U} \rightarrow {}^{234}_{90}\text{Th} + {}^{4}_{2}\text{He} + {}^{0}_{0}\gamma$	α, γ
$^{234}_{90}\text{Th} \rightarrow {}^{234}_{91}\text{Pa} + {}^{0}_{-1}\text{e} + {}^{0}_{0}\gamma$	β, γ
$^{234}_{91}\text{Pa} \rightarrow {}^{234}_{92}\text{U} + {}^{0}_{-1}\text{e} + {}^{0}_{0}\gamma$	β, γ
$^{234}_{92}\text{U} \rightarrow {}^{230}_{90}\text{Th} + {}^{4}_{2}\text{He} + {}^{0}_{0}\gamma$	α, γ
$^{230}_{90}\text{Th} \rightarrow {}^{226}_{88}\text{Ra} + {}^{4}_{2}\text{He} + {}^{0}_{0}\gamma$	α, γ
$^{226}_{88}\text{Ra} \rightarrow {}^{222}_{86}\text{Rn} + {}^{4}_{2}\text{He}$	α
$^{222}_{86}\text{Rn} \rightarrow {}^{218}_{84}\text{Po} + {}^{4}_{2}\text{He}$	α

7.4 Detecting radioactivity

You are not required to know the details of the following detectors of radioactive emissions, but you will certainly use one or more of the devices in your practical work. Furthermore, some of these detectors are based on the ionising properties of the particles or radiation, and it is interesting to apply our previous knowledge to explain how they work.

The Geiger counter

Figure 7.4 illustrates a Geiger-Müller tube with a scaler connected to it. When radiation enters the window, it creates ion pairs in the gas in the tube. These charged particles, and particularly the electrons, are accelerated by the potential difference between the central wire anode and the cylindrical cathode. These accelerated particles then cause further ionisation. The result of this continuous process is described as an **avalanche effect**. That is, the entry of one particle into the tube and the production of one ion pair results in very large numbers of electrons and ions arriving at the anode and cathode respectively. This gives a pulse of charge which is amplified and counted by the scaler or ratemeter. (A scaler measures the total count of pulses in the tube during the time that the scaler is operating. A ratemeter continuously monitors the number of counts per second.) Once the pulse has been registered, the charges are removed from the gas in readiness for further radiation entering the tube.

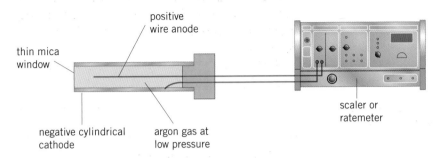

Figure 7.4 Geiger-Müller tube and scaler

Photographic plates

Figure 7.5 Film badge dosimeter

When a radioactive emission strikes a photographic film, the film reacts as if it had been exposed to a small amount of visible light. When the film is developed, fogging or blackening is seen. This fogging can be used to detect not only the presence of radioactivity, but also the intensity of the radiation.

Figure 7.5 shows a film badge dosimeter. It contains a piece of photographic film which becomes fogged when exposed to radiation. Workers who are at risk from radiation wear such badges to gauge the type and dose of radiation to which they have been exposed. The radiation passes through different filters before reaching the film. Consequently the type of radiation, as well as the quantity, can be assessed.

The scintillation counter

Early workers with radioactive materials used glass screens coated with zinc sulphide to detect radiation. When radiation is incident on the zinc sulphide, it emits a tiny pulse of light called a **scintillation**. The rate at which these pulses are emitted indicates the intensity of the radiation.

The early researchers worked in darkened rooms, observing the zinc sulphide screen by eye through a microscope and counting the number of flashes of light occurring in a certain time. Now a scintillation counter is used (Figure 7.6).

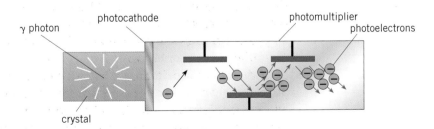

Figure 7.6 Scintillation counter

Often a scintillator crystal is used instead of a zinc sulphide screen. The crystal is mounted close to a device known as a photomultiplier, a vacuum-tube device which uses the principle of photoelectric emission. Flashes of light cause the emission of photoelectrons from the negative electrode of the photomultiplier. The photoelectric current is amplified inside the tube. The output electrode is connected to a scaler or ratemeter, as with the Geiger-Müller tube.

Background radiation

Radioactivity is a natural phenomenon. Rocks such as granite contain small amounts of radioactive nuclides, some foods we eat emit radiation, and even our bodies are naturally radioactive. Although the atmosphere provides life on Earth with some shielding, there is, nevertheless, some radiation from outer space (cosmic radiation). In addition to this natural radioactivity, we are exposed to radiation from man-made sources. These are found in medicine, in fallout from nuclear explosions, and in leaks from nuclear power stations. The sum of all this radiation is known as **background radiation**. Figure 7.7 indicates the relative proportions of background radiation coming from various sources.

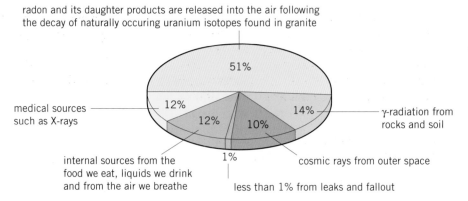

Figure 7.7 Sources of background radiation

In carrying out experiments with radioactive sources, it is important to take account of background radiation. In order to determine the count-rate due to the radioactive source, the background count-rate must be subtracted from the total measured count-rate. Allowance for background radiation gives the corrected count-rate.

The spontaneous and random nature of radioactive decay

Detection of the count-rate of radioactive sources shows that the emission of radiation is both *spontaneous* and *random*. It is a spontaneous process because it is not affected by any external factors, such as temperature or pressure. Decay is random in that it is not possible to predict which nucleus in a sample will decay next. There is, however, a constant probability (or chance) that a nucleus will decay in any fixed period of time. We will look at this in more detail in Section 7.5.

Sections 7.2–7.4 Summary

- An α-particle is a helium nucleus (two protons + two neutrons).
- A β-particle is a fast-moving electron.
- γ-radiation consists of short-wavelength electromagnetic waves.
- Nuclear notation: $^{\text{nucleon number}}_{\text{proton number}}X$, where X is the symbol for the chemical element.
- In nuclear notation the emissions are represented as: α-particle $^4_2$He; β-particle $^0_{-1}$e; γ-radiation $^0_0\gamma$
- α-emission reduces the nucleon number of the parent nucleus by 4, and reduces the proton number by 2.
- β-emission causes no change to the nucleon number of the parent nucleus, and increases the proton number by 1.
- γ-emission causes no change to nucleon number or proton number of the parent nucleus.
- Radioactive decay is a spontaneous, random process.

Sections 7.2–7.4 Questions

1 You are provided with a radioactive source which could be emitting α- or β-particles, or γ-radiation, or any combination of these. Describe a simple experiment, based on their relative penetrating qualities, you might carry out to determine the nature of the radiation(s) being emitted.

2 Explain the changes that take place to the nucleus of an atom when it emits:
(a) an α-particle,
(b) a β-particle,
(c) γ-radiation.

3 Complete the following radioactive series.

$$^a_b X \rightarrow\ ^?_? Y + ^4_2 He$$
$$^?_? Y \rightarrow\ ^?_? Z + ^0_{-1} e$$
$$^?_? Z \rightarrow\ ^?_? Z + ^?_? ?$$

7.5 Random decay

At the end of Sections 7.5–7.6 you should be able to:

- define half-life
- define the terms activity and decay constant and recall and solve problems using $A = \lambda N$
- infer and sketch the exponential nature of radioactive decay and solve problems using the relationship $x = x_0 \exp(-\lambda t)$ where x could represent activity, number of undecayed particles or received count-rate
- solve problems using the relation $\lambda = 0.693/t_{\frac{1}{2}}$
- describe the use of a radioactive isotope in a smoke detector
- describe the technique of radioactive dating (carbon-dating).

We now look at some of the consequences of the random nature of radioactive decay. If six dice are thrown simultaneously (Figure 7.8), it is likely that one of them will show a six. If twelve dice are thrown, it is likely that two of them will show a six, and so on. While it is possible to predict the likely number of sixes that will be thrown, it is impossible to say which of the dice will actually show a six. We describe this situation by saying that the throwing of a six is a **random process**.

Figure 7.8 The dice experiment

In an experiment similar to the one just described, some students throw a large number of dice (say 6000). Each time a six is thrown, that die is removed. The results for the number of dice remaining after each throw are shown in Table 7.4.

Table 7.4 Results of dice-throwing experiment

number of throws	number of dice remaining	number of dice removed
0	6000	
1	5000	1000
2	4173	827
3	3477	696
4	2897	580
5	2414	483
6	2012	402
7	1677	335
8	1397	280

Figure 7.9 is a graph of the number of dice remaining against the number of throws. This kind of graph is called a **decay curve**. The rate at which dice are removed is not linear, but there is a pattern. After between 3 and 4 throws, the number of dice remaining has halved. Reading values from the graph shows that approximately 3.8 throws would be required to halve the number of dice. After another 3.8 throws the number has halved again, and so on.

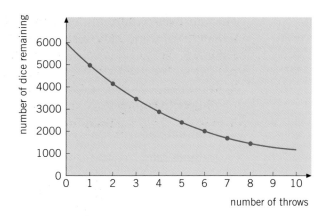

Figure 7.9

We have met this type of curve before, in talking about the decay of current in a circuit containing a capacitor and a resistor (Chapter 5).

We can apply the dice experiment to model radioactive decay. The 6000 dice represent radioactive nuclei. To score a six represents radioactive emission. All dice scoring six are removed, because once a nucleus has undergone radioactive decay, it is no longer available for further decay. Thus, we can describe how rapidly a sample of radioactive material will decay.

A graph of the number of undecayed nuclei in a sample against time has the typical decay curve shape shown in Figure 7.10. It is not possible to state how long the entire sample will take to decay (its 'life'). However, after 3 minutes the number of undecayed nuclei in the sample has halved. After a further 3 minutes, the number of undecayed atoms has halved again. We describe this situation by saying that this radioactive nuclide has a **half-life** $t_{\frac{1}{2}}$ of 3 minutes.

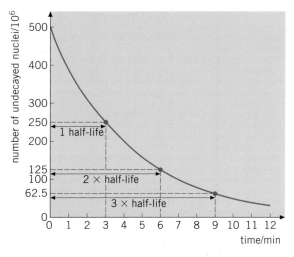

Figure 7.10 Radioactive decay curve

The half-life of a radioactive nuclide is the time taken for the number of undecayed nuclei to be reduced to half its original number.

The half-lives of different nuclides have a very wide range of values. Examples of some radioactive nuclides and their half-lives are given in Table 7.5.

Table 7.5 Examples of half-lives

radioactive nuclide	half-life
uranium-238	4.5×10^9 years
radium-226	1.6×10^3 years
radon-222	3.8 days
francium-221	4.8 minutes
astatine-217	0.03 seconds

We shall see later that half-life may also be expressed in terms of the activity of the material.

Note that half-life is similar to the time constant, which we met in Chapter 5. Both half-life and the time constant relate to the time taken for a certain fraction of the quantity concerned to decay. For the half-life, that fraction is 1/2. For the time constant, the fraction is 1/e, or about 1/2.7.

If you measure the count-rate from a nuclide with a very long half-life, you might expect to obtain a constant value. In fact, the reading of the ratemeter fluctuates about an average value. This demonstrates the random nature of radioactive decay.

Example

The half-life of francium-221 is 4.8 minutes. Calculate the fraction of a sample of francium-221 remaining undecayed after a time of 14.4 minutes.

The half-life of francium-221 is 4.8 min, so after 4.8 min half of the sample will remain undecayed. After two half-lives (9.6 min), $0.5 \times 0.5 = 0.25$ of the sample will remain undecayed. After three half-lives (14.4 min), $0.5 \times 0.25 = 0.125$ will remain undecayed. So the fraction remaining undecayed is **0.125** or **1/8**.

Now it's your turn

Using the half-life values given in Table 7.5, calculate:
(a) the fraction of a sample of uranium-238 remaining undecayed after 9.0×10^9 years,
(b) the fraction of a sample of astatine remaining undecayed after 0.30 s,
(c) the fraction of a sample of radium-226 that has decayed after 3200 years,
(d) the fraction of a sample of radon-222 that has decayed after 15.2 days.

Applications of the use of radioactivity

The smoke detector

The current UK Building Regulations require all new buildings to be fitted with a smoke detector and alarm. It works by the ionisation of air caused by a small α-particle emitter with a long half-life. The α-particles pass between two electrodes in an ionisation chamber open to the air of the room. Ionisation of the air produces a small, constant current between the electrodes. However, if the air contains more than a certain concentration of smoke, the smoke particles absorb the α-particles and reduce the current. When the current falls below a certain preset limit, an audible alarm is triggered.

Why use α-particles, and not, for example, a β-particle source? α-particles are much more efficient at ionising air than β-particles, and also have low penetration. They are stopped by the casing of the detector and thus offer no hazard to people in the room. A source that is often used is radioactive americium-241 which has a half-life of over 400 years, another advantage. The mass of the source is very small, less than a milligram.

Figure 7.11 Ionisation smoke alarm

Radioactive dating (carbon-dating)

The age of ancient objects made of once-living matter, such as wood, can be determined using a technique known as *radioactive dating*. The method makes use of the natural radioactivity of carbon-14.

All living plants absorb carbon dioxide from atmospheric air and use it in their growth, forming organic molecules which contain carbon atoms. Of the carbon atoms in the air, by far the largest number are the carbon-12 isotope, which is not radioactive, but a very small fraction, of the order of 10^{-12}, is the radioactive isotope carbon-14. This fraction has remained constant over many thousands of years. Carbon-14 decays with a half-life of 5730 years.

When the plant dies, it ceases to take up carbon dioxide from the air. The carbon-14 in it decays, but the carbon-12 does not; thus, the ratio of carbon-14 to carbon-12 in a sample of plant material can be used to estimate how long ago the plant died. This principle can also be applied to wooden artefacts, for example the wooden handles of tools or weapons, or to ashes from fires in ancient settlements. Suppose the handle of an ancient tool shows a carbon-14 to carbon-12 ratio of only half of what the ratio is in living trees. An archaeologist can deduce that the branch from which the handle was made was cut from its tree one half-life of carbon-14 ago, that is the tool is about 5700 years old.

Because of the difficulty of determining very low levels of activity accurately, carbon-dating can be used only over a period of up to about ten half-lives, that is up to about 60 000 years. But, on the geological timescale, it has been possible to date the oldest rocks on Earth to about 4×10^9 years. Here, the radioactive uranium-238 isotope has been used. Uranium-238 has a half-life of about 4.5×10^9 years. The concentration of uranium-238 in a rock relative to its daughter nuclei becomes fixed when the molten materials making up the rock solidify, effectively starting the radioactive dating clock. The earliest fossil-bearing rocks are about 3×10^9 years old, helping us to estimate the date at which life first appeared on Earth.

7.6 Mathematical descriptions of radioactive decay

Activity and decay constant

As we saw in the dice experiment, increasing the number of dice increases the number of sixes that appear with each throw. Similarly, if we investigate the decay of a sample of radioactive material we find that the greater the number of radioactive nuclei in the sample the greater the rate of decay. This can be described mathematically as $dN/dt \propto N$. This gives

$$\frac{dN}{dt} = -\lambda N$$

N is the number of undecayed atoms in the sample. dN/dt is the rate at which the number of nuclei in the sample is changing, so $-dN/dt$ represents the rate of decay. $-dN/dt$ is known as the **activity** A of the source, and is measured in **becquerels**.

The activity of a radioactive source is the number of nuclear decays produced per unit time in the source. Activity is measured in becquerels (Bq), where 1 becquerel is 1 decay per second.

$$1 \text{ Bq} = 1 \text{ s}^{-1}$$

Combining $A = -\mathrm{d}N/\mathrm{d}t$ and $\mathrm{d}N/\mathrm{d}t = -\lambda N$, we have

$$A = \lambda N$$

λ is a constant of proportionality known as the **decay constant**. It has the units s^{-1}, yr^{-1} and so on.

λ is defined as the probability per unit time that a nucleus will undergo decay.

This is an important equation because it relates a quantity we can measure ($-\mathrm{d}N/\mathrm{d}t$, the rate of decay or activity) to a quantity which cannot in practice be determined (N, the number of undecayed nuclei). We shall see later that the decay constant λ is directly related to the half-life $t_{\frac{1}{2}}$, which can be obtained by experiment. This opens the way to calculating the number of undecayed nuclei in a sample. Trying to count nuclei when they are decaying is similar to counting sheep in a field while some are escaping through a gap in the hedge!

Example

Calculate the number of phosphorus-32 nuclei in a sample which has an activity of 5.0×10^6 Bq.
(decay constant of phosphorus-32 = $5.6 \times 10^{-7} \text{ s}^{-1}$)

From $\mathrm{d}N/\mathrm{d}t = -\lambda N$, $N = (-\mathrm{d}N/\mathrm{d}t)/\lambda = -5.0 \times 10^6 / 5.6 \times 10^{-7} = -8.9 \times 10^{12}$. The minus sign in this answer arises because $\mathrm{d}N/\mathrm{d}t$ is the rate of decay. The quantity measured by a ratemeter is the rate of decay, and so should be negative, but it is always displayed as a positive quantity. Similarly, activities are always quoted as positive. So don't be worried about discarding the minus sign here! The number of phosphorus-32 nuclei = **8.9×10^{12}**.

Note: In this type of calculation, because the activity is measured in becquerel (s^{-1}), the decay constant λ must be measured in consistent units. If λ had been quoted as 4.2 day^{-1}, it would have been necessary to convert to s^{-1}.

Now it's your turn

Calculate the activity of the following samples of radioactive materials:
(a) 6.7×10^{21} atoms of strontium-90,
   (decay constant of strontium-90 = $8.3 \times 10^{-10} \text{ s}^{-1}$)
(b) 2.0 mg of uranium-238. 0.238 kg of uranium-238 contains 6.0×10^{23} atoms.
   (decay constant of uranium-238 = $5.0 \times 10^{-13} \text{ s}^{-1}$)

To solve the equation $dN/dt = -\lambda N$ requires mathematics beyond the scope of A/AS Physics. However, it is important to know the solution, in order to find the variation with time of the number of nuclei remaining in the sample. The solution is

$$N = N_0 e^{-\lambda t} \quad \text{or} \quad N = N_0 \exp(-\lambda t)$$

where N_0 is the initial number of undecayed nuclei in the sample, and N is the number of undecayed nuclei at time t. This equation is of exactly the same form as the one used to describe the decay of current in a circuit containing a capacitor and a resistor. Note that the decay constant λ is analogous to the reciprocal of the time constant, that is to $1/CR$.

The equation represents an **exponential** decay. The decay curve of N against t is as shown in Figure 7.10. Since activity A is proportional to N ($A = \lambda N$), the curve of A against t is the same shape, and we can write

$$A = A_0 e^{-\lambda t}$$

Example

A sample of phosphorus-32 contains 8.6×10^{12} nuclei at time $t = 0$. The decay constant of phosphorus-32 is 4.8×10^{-2} day^{-1}. Calculate the number of undecayed phosphorus-32 nuclei in the sample after 10 days.

From $N = N_0 e^{-\lambda t}$, we have $N = 8.6 \times 10^{12} \times e^{-0.048 \times 10}$, so $N = \mathbf{5.3 \times 10^{12}}$. (Again, it is important to measure λ and t in consistent units. Here λ is in day^{-1} and t is in days, so there is no problem.)

Now it's your turn

A sample of phosphorus-32 contains 8.6×10^{12} nuclei at time $t = 0$. The decay constant of phosphorus-32 is 4.8×10^{-2} day^{-1}. Calculate the number of undecayed phosphorus-32 nuclei in the sample after:
(a) 20 days,
(b) 40 days.

Decay constant and half-life

Using the equation $N = N_0 e^{-\lambda t}$, we can derive an equation which relates the half-life to the decay constant. For any radioactive nuclide, the number of undecayed nuclei after one half-life is, by the definition of half-life, equal to $N_0/2$, where N_0 is the original number of undecayed nuclei. Using the radioactive decay equation

$$N = N_0 e^{-\lambda t}$$

we have, at time $t = t_{\frac{1}{2}}$

$$N = N_0 e^{-\lambda t_{\frac{1}{2}}}$$

and, dividing each side of the equation by N_0,

$$\frac{N}{N_0} = e^{-\lambda t_{\frac{1}{2}}} \quad \text{so} = \frac{1}{2} = e^{-\lambda t_{\frac{1}{2}}}$$

or

$$2 = e^{\lambda t_{\frac{1}{2}}}$$

Taking natural logarithms of both sides,

$$\ln 2 = \lambda t_{\frac{1}{2}}$$

So that

$$t_{\frac{1}{2}} = \frac{\ln 2}{\lambda} \quad \text{or} \quad t_{\frac{1}{2}} = \frac{0.693}{\lambda}$$

Example

Calculate the half-life of radium-226, which has a decay constant of $1.42 \times 10^{-11}\,\text{s}^{-1}$.

Using $t_{\frac{1}{2}} = 0.693/\lambda$, we have $t_{\frac{1}{2}} = 0.693/1.42 \times 10^{-11} = \textbf{4.88} \times \textbf{10}^{\textbf{10}}\textbf{s}$.

Now it's your turn

1 Calculate the half-lives of the following radioactive nuclides:
   (a) bismuth-214, which has a decay constant of $4.3 \times 10^3\,\text{s}^{-1}$,
   (b) carbon-14, which has a decay constant of $4.1 \times 10^{-12}\,\text{s}^{-1}$.

2 Calculate the decay constants of the following radioactive nuclides:
   (a) helium-5, which has a half-life of $6.0 \times 10^{-20}\,\text{s}$,
   (b) sodium-24, which has a half-life of $15\,\text{h}$.

Sections 7.5–7.6 Summary

- The half-life $t_{\frac{1}{2}}$ of a radioactive nuclide is the time taken for the number of undecayed nuclei to be reduced to half the original number.
- The activity of a radioactive source is the number of decays per unit time. The unit of activity is the becquerel (Bq); 1 becquerel = 1 decay per second.
- The activity dN/dt of a source is related to the number N of undecayed nuclei by the equation:

$$dN/dt = -\lambda N$$

where λ is the decay constant.

> ■ The decay constant is defined as the probability of decay per unit time of a nucleus.
> ■ The number N of undecayed nuclei in a radioactive sample at time t is given by the equation:
>
> $$N = N_0\, e^{-\lambda t}$$
>
> where N_0 is the number of undecayed nuclei at time $t = 0$.
> ■ The half-life $t_{\frac{1}{2}}$ and the decay constant λ are related by the equation:
>
> $$t_{\frac{1}{2}} = 0.693/\lambda$$

Sections 7.5–7.6 Questions

1 The table below shows the variation with time t of the activity of a sample of a radioactive nuclide X. The average background count during the experiment was $36\,\text{min}^{-1}$.
  (a) Plot a graph to show the variation with time of the corrected count-rate.
  (b) Use the graph to determine the half-life of the nuclide X.

t/hour	activity/min^{-1}	t/hour	activity/min^{-1}
0	854	6	448
1	752	7	396
2	688	8	362
3	576	9	334
4	544	10	284
5	486		

2 A radioactive source with a half-life of 10 days has an initial activity of $2.0 \times 10^{10}\,\text{Bq}$. Calculate the activity of this source after 30 days.

3 The half-lives of two radioactive nuclides X and Y are 40 s and 70 s respectively. Initially, samples of each contain 2.0×10^7 undecayed nuclei.
  (a) Calculate the initial activity of each sample.
  (b) Draw, on the same axes, two lines to represent the decay of these samples.

4 Calculate the mass of caesium-137 that has an activity of $4.0 \times 10^5\,\text{Bq}$. 0.137 kg of caesium-137 contains 6.0×10^{23} atoms. The half-life of caesium-137 is 30 years.

5 (a) The activity of a radioactive source X falls from $5.0 \times 10^{10}\,\text{Bq}$ to $1.0 \times 10^{10}\,\text{Bq}$ in 5.0 hours. Calculate the half-life.
  (b) The activity of a certain mass of carbon-14 is $5.00 \times 10^9\,\text{Bq}$. The half-life of carbon-14 is 5570 years. Calculate the number of carbon-14 nuclei in the sample.

6 A sample of wood from the remains of a palisade on an ancient Irish crannog (a fortified dwelling built on an artificial island in a lake) contains 170 g of carbon and has an activity of 36 decays per second, due to the decay of the radioactive carbon-14 in the sample. The half-life of carbon-14 is 5730 years. A mass of 12 g of carbon contains 6.0×10^{23} carbon atoms. In living trees the ratio of carbon-14 to carbon-12 atoms is about 1.3×10^{-12}. Estimate the age of the palisade.

7.7 Mass defect

At the end of Sections 7.7–7.11 you should be able to:
- show an appreciation of the association between energy and mass as represented by $E = mc^2$ and recall and solve problems using this relationship
- sketch the variation of binding energy per nucleon with nucleon number
- explain the relevance of binding energy per nucleon to nuclear fusion and to nuclear fission.

At a nuclear level, the masses we deal with are so small that it would be very clumsy to measure them in kilograms. Instead, we measure the masses of nuclei and nucleons in **atomic mass units (u)**.

One atomic mass unit (1 u) is defined as being equal to one-twelfth of the mass of a carbon-12 atom. 1 u is equal to 1.66×10^{27} kg.

Using this scale of measurement, to six decimal places, we have

proton mass $m_p = 1.007276$ u

neutron mass $m_n = 1.008665$ u

electron mass $m_e = 0.000549$ u

Because all atoms and nuclei are made up of protons, neutrons and electrons, we should be able to use these figures to calculate the mass of any atom or nucleus. For example, the mass of a helium-4 nucleus, consisting of two protons and two neutrons, should be

$(2 \times 1.007276) + (2 \times 1.008665) = 4.031882$ u

However, the actual mass of a helium nucleus is 4.001508 u.

The difference between the expected mass and the actual mass of a nucleus is called the **mass defect** of the nucleus. In the case of the helium-4 nucleus, the mass defect is $4.031882 - 4.001508 = 0.030374$ u.

The mass defect of a nucleus is the difference between the total mass of the separate nucleons and the combined mass of the nucleus.

> ## Example
>
> Calculate the mass defect for a carbon-14 ($^{14}_{6}$C) nucleus. The measured mass is 14.003240 u.
>
> The nucleus contains 6 protons and 8 neutrons, of total mass $(6 \times 1.007276) + (8 \times 1.008665) = 14.112976$ u. The mass defect is $14.112976 - 14.003240 = $ **0.109736 u**.
>
> ## Now it's your turn
>
> Calculate the mass defect for a nitrogen-14 ($^{14}_{6}$N) nucleus. The measured mass is 14.003070 u.

7.8 Mass–energy equivalence

In 1905, Albert Einstein proposed that there is an equivalence between mass and energy. The relationship between energy E and mass m is

$$E = mc^2$$

where c is the speed of light. E is measured in joules, m in kilograms and c in metres per second.

Using this relation, we can calculate that 1.0 kg of matter is equivalent to $1.0 \times (3.0 \times 10^8)^2 = 9.0 \times 10^{16}$ J.

The mass defect of the helium nucleus, calculated previously as 0.030374 u, is equivalent to $0.030374 \times 1.66 \times 10^{-27} \times (3.00 \times 10^8)^2 = 4.54 \times 10^{-12}$ J. (Note that the mass in u must be converted to kg by multiplying by 1.66×10^{-27}.)

The joule is an inconveniently large unit to use for nuclear calculations. A more convenient energy unit is the mega electronvolt (MeV), as many energy changes that take place in the nucleus are of the order of several MeV. One mega electronvolt is the energy gained by one electron when it is accelerated through a potential difference of one million volts.

Since *electrical energy = charge × potential difference*, and the electron charge is 1.60×10^{-19} C,

$$1\,\text{MeV} = 1.60 \times 10^{-19} \times 1.00 \times 10^6$$

or

$$1\,\text{MeV} = 1.60 \times 10^{-13}\,\text{J}$$

The energy equivalent of the mass defect of the helium nucleus is thus $4.54 \times 10^{-12}/1.60 \times 10^{-13} = 28.4$ MeV.

If mass is measured in u and energy in MeV,

1 u is the equivalent of 931 MeV

7.9 Binding energy

Within the nucleus there are strong forces which bind the protons and neutrons together. To completely separate all these nucleons requires energy. This energy is referred to as the **binding energy** of the nucleus. Stable nuclei, those which have little or no tendency to disintegrate, have large binding energies. Less stable nuclei have smaller binding energies.

Similarly, when protons and neutrons are joined together to form a nucleus, this binding energy must be released. The binding energy is the energy equivalent of the mass defect.

We have seen that the binding energy of the helium-4 nucleus is 28.4 MeV. 28.4 MeV of energy is required to separate, to infinity, the two protons and two neutrons of this nucleus.

> Binding energy is the energy equivalent of the mass defect of a nucleus. It is the energy required to separate to infinity all the nucleons of a nucleus.

Example

Calculate the binding energy, in MeV, of a carbon-14 nucleus with a mass defect of 0.109736 u.

Using the equivalence 1 u = 931 MeV, 0.109736 u is equivalent to 102 MeV. Since the binding energy is the energy equivalent of the mass defect, the binding energy = **102 MeV**.

Now it's your turn

Calculate the binding energy, in MeV, of a nitrogen-14 nucleus with a mass defect of 0.108517 u.

Stability of nuclei

A stable nucleus is one which has a very low probability of decay. Less stable nuclei are more likely to disintegrate. A useful measure of stability is the **binding energy per nucleon** of the nucleons in the nucleus.

> Binding energy per nucleon is defined as the total energy needed to completely separate all the nucleons in a nucleus divided by the number of nucleons in the nucleus.

Figure 7.12 shows the variation with nucleon number of the binding energy per nucleon for different nuclides. The most stable nuclides are those with the highest binding energy per nucleon; that is, those near point B on the graph. Iron is one of these stable nuclides. Typically, very stable nuclides have binding energies per nucleon of about 8 MeV.

Light nuclei, between A and B on the graph, may combine or fuse to form larger nuclei with larger binding energies per nucleon. This process is called **nuclear fusion**. For the process to take place, conditions of very high temperature and pressure are required, such as in stars like the Sun.

Heavy nuclei, between B and C on the graph, when bombarded with neutrons, may break into two smaller nuclei, again with larger binding energy per nucleon values. This process is called **nuclear fission**.

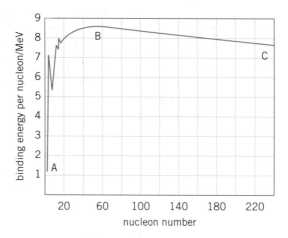

Figure 7.12 Binding energy per nucleon, against nucleon number

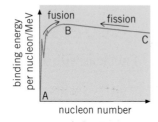

Figure 7.13

Figure 7.13 highlights nuclides which may undergo fusion (blue part of curve) or fission (red part of curve) in order to increase their binding energy per nucleon.

When nuclear fusion or fission takes place, the nucleon numbers of the nuclei involved change. A higher binding energy per nucleon is achieved, and this is accompanied by a *release* of energy. This release of energy during fission reactions is how the present generation of nuclear power stations produce electricity.

Example

The binding energy of a helium-4 nucleus is 28.4 MeV. Calculate the binding energy per nucleon.

The helium-4 nucleus has 4 nucleons. The binding energy per nucleon is thus 28.4/4 = **7.1 MeV per nucleon**.

Now it's your turn

The binding energy of a carbon-14 nucleus is 102 MeV. Calculate the binding energy per nucleon.

7.10 Nuclear fission

Within the nucleus of an atom the nucleons experience both attractive and repulsive forces. The attractive force is called the **strong nuclear force**. This acts like a 'nuclear glue' to hold the nucleons together. The repulsive forces are the electric (Coulomb-law) forces between the positively charged protons. Gravitational forces of attraction exist, but are negligible in comparison to the other forces. We will justify this later in section 7.12. Stable nuclei have much larger attractive forces than repulsive forces. Stable nuclides generally have approximately the same number of neutrons and protons in the nucleus; that is, the neutron-to-proton ratio is close to one. In heavy nuclei such as uranium and plutonium there are far more neutrons than protons, giving a neutron-to-proton ratio of more than one. For example, uranium-235 has 92 protons and 143 neutrons, giving a neutron-to-proton ratio of 1.55. This leads to a much lower binding energy per nucleon compared with iron, and such nuclides are less stable. Any further increase in the number of neutrons in such nuclei is likely to cause the nucleus to undergo **nuclear fission**.

Nuclear fission is the splitting of a heavy nucleus into two lighter nuclei of approximately the same mass.

When a uranium-235 nucleus absorbs a neutron, it becomes unstable and splits into two lighter, more stable nuclei. The most likely nuclear reaction is

$$\ce{^{235}_{92}U} + \ce{^{1}_{0}n} \rightarrow \ce{^{236}_{92}U} \rightarrow \ce{^{141}_{56}Ba} + \ce{^{92}_{36}Kr} + 3\ce{^{1}_{0}n} + \text{energy}$$

This process is called *induced* nuclear fission, because it is started by the capture of a neutron by the uranium nucleus.

Each of the fission reactions described by this equation results in the release of three neutrons. Other possible fission reactions release two or three neutrons. If these neutrons are absorbed by other uranium-235 nuclei, these too may become unstable and undergo fission, thereby releasing even more neutrons. The reaction is described as being a **chain reaction** which is accelerating. This is illustrated in Figure 7.14. If this type of reaction continues uncontrolled, a great deal of energy is released in a short time, and a nuclear explosion results.

If the number of neutrons which take part in the chain reaction is controlled so that the number of fissions per unit time is constant, rather than increasing, the rate of release of energy can be controlled. This situation is illustrated in Figure 7.15. These conditions apply in the reactor of a modern nuclear power station, where some of the neutrons released in fission reactions are absorbed by control rods in order to limit the rate of fission reactions.

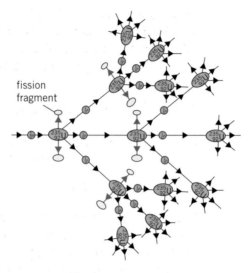

fission
fragment

Figure 7.14 Accelerating chain reaction

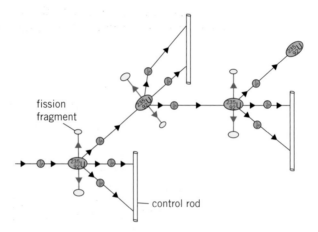

fission
fragment

control rod

Figure 7.15 Controlled chain reaction. Two of the neutrons produced by fission are absorbed by control rods. The third neutron induces further nuclear fission.

Peaceful and destructive uses of nuclear fission

The immense amount of energy produced in fission reactions at once suggests an application in the generation of electric power for industrial and domestic use. Fission reactors, to be described in the next section, are generally described as *power* reactors if the kinetic energy of the neutrons is harnessed to produce steam to drive turbine generators. Because nuclear fission produces so many neutrons, *research* reactors are also constructed to provide intense beams of neutrons for use as projectiles in nuclear reactions. These reactions may also be used to prepare radioactive nuclides not naturally found, including those used as tracers and for medical treatment.

We have summarised some of the benign uses of nuclear fission. However, its first use was a military one, causing immense destruction of cities and widespread fatalities among the civilian population. The first fission bomb (miscalled an 'atomic' bomb) was constructed by a team of Allied scientists working at Los Alamos, New Mexico, and tested in the New Mexico desert in July 1945. In August 1945 fission bombs were dropped on Hiroshima and Nagasaki, bringing the Second World War to an end.

Nuclear fission reactor

There are several possible fission reactions which may occur in a nuclear reactor. Most are represented by the general equation

$$^{235}_{92}U + {}^{1}_{0}n \rightarrow {}^{236}_{92}U \rightarrow 2 \text{ new nuclides} + 2 \text{ or } 3 \text{ neutrons} + \text{energy}$$

The nuclides formed in this reaction are called **fission fragments**.

Figure 7.16 illustrates a pressurised water-cooled reactor (PWR). Inside the reactor, the uranium-235 is placed in long fuel rods which are surrounded by heavy water (water in which deuterium atoms, hydrogen-2, take the place of the normal hydrogen-1 atoms) or graphite. The role of these materials is to slow down the neutrons released during fission. This is important because slow-moving neutrons are more readily absorbed by uranium-235. Materials such as heavy water or graphite which slow down the neutrons are called **moderators**.

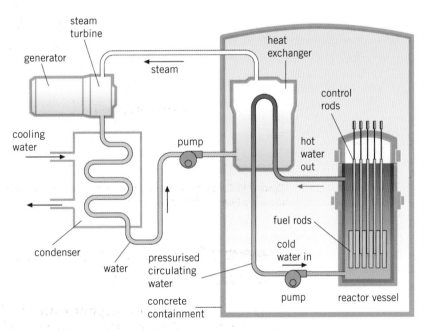

Figure 7.16 Pressurised water-cooled reactor

The number of neutrons available to trigger further fission reactions is controlled using **control rods** of boron or cadmium. These substances readily absorb neutrons. By pushing the rods further into the reactor core, the number of neutrons absorbed increases and the rate of fission reactions decreases. If the rods are pulled out a little, fewer neutrons are absorbed and the rate of fissioning increases. Ideally, the rods should be adjusted so that the fission reaction proceeds at a constant rate; that is, it neither accelerates nor decelerates. The energy released during fission reactions is seen mainly as the kinetic energy of the resulting lighter nuclei and neutrons (the fission fragments). When the fission fragments are slowed down, the heat energy produced is removed from the reactor by a circulating coolant. This heat energy is then used to produce high temperature steam in a heat exchanger. The steam is used to drive turbine generators.

There are two main types of nuclear reactor used for the commercial generation of electricity. These are the pressurised water-cooled reactor (PWR) and the advanced gas-cooled reactor (AGR).

In the PWR, water circulates through the reactor vessel at a high pressure, typically 150 atmospheres. At such pressure the water does not boil, even though its temperature may be as high as 300 °C. This super-heated water passes into a heat exchanger through a series of pipes surrounded by water at much lower pressure. The low-pressure water absorbs heat energy and changes into steam, which then drives the turbines and electricity generators.

In AGR reactors, helium gas at a temperature of about 600 °C is the coolant. Again, energy is transferred from the core to water to a heat exchanger in order to produce the steam to drive the turbines.

Environmental hazards of nuclear fission

When a fission bomb explodes, many highly radioactive fission fragments are released into the atmosphere, causing what is known as *nuclear fallout*. The movement of air in the atmosphere causes a circulation of this radioactive material round the Earth. It eventually falls to the ground, settling on crops and other vegetation, and is absorbed so that it eventually enters the food chain. This produces a much more serious risk to humans than the fallout which settles on clothing or body; α- and β-radiation is likely to be absorbed by clothing and the outer, dead, layer of skin so that it does not enter the body and cause damage to interior organs. But radioactive materials ingested with food bring the isotopes into direct contact with living cells. A particularly dangerous isotope is strontium-90, which, being chemically very similar to calcium, becomes concentrated in bone. There it causes bone cancer and the destruction of bone marrow.

The fallout hazard was not fully recognised until 1963. Then an international nuclear testing treaty was signed, banning the testing of weapons in the atmosphere.

The peaceful use of nuclear fission also presents many hazards. Even disregarding the accidental discharge of radioactive fission fragments into the atmosphere, such as occurred in incidents at Three Mile Island in the USA in 1979 and at Chernobyl in the Ukraine in 1986, there is the question of the routine disposal of the fission fragments and of spent fuel. The spent fuel can be reprocessed, but other waste must be stored underground in strong steel containers. Nuclear fission reactors have a lifetime of only about 30 years because of the build-up of radioactivity and the weakening of the structural materials, and there are particular risks when it comes to decommissioning a reactor.

7.11 Nuclear fusion

Most of the energy on Earth comes from the Sun, where it is produced by nuclear fusion reactions. Light nuclei, such as isotopes of hydrogen, join together to produce heavier, more stable nuclei, and in doing so release energy. One of these fusion reactions is

$$^2_1H + ^2_1H \rightarrow ^3_2He + ^1_0n + \text{energy}$$

From the binding energy per nucleon curve (Figure 7.12) we see that the binding energy per nucleon for light nuclei, such as hydrogen, is low. But if two light nuclei are made to fuse together, they may form a new heavier nucleus which has a higher binding energy per nucleon. It will be more stable than the two lighter nuclei from which it was formed. Because of this difference in stability, a fusion reaction such as this will release energy.

Although fusion reactions are the source of solar energy, we are at present unable to duplicate this reaction in a controlled manner on Earth. This is because the nuclei involved in fusion have to be brought very close together, and conditions of extremely high temperature and pressure, similar to those found at the centre of the Sun, are required. Reactions requiring these conditions are called **thermonuclear reactions**. Some fusion reactions involving hydrogen isotopes have been made to work in the Joint European Torus (JET), although not yet in a controlled, sustainable manner. In 2006, an international consortium agreed to undertake the ITER (International Tokamak Engineering Research) project, which is designed to produce up to 500 MW of fusion power sustained for over 400 seconds by the fusion of a ^2H–^3H mixture. Construction on a site in southern France will take 10 years.

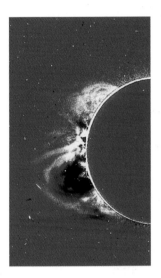

Figure 7.17 Most of our energy on Earth comes from the Sun. It is produced by nuclear fusion reactions. Light nuclei such as isotopes of hydrogen join together to produce heavier, more stable nuclei. This photo shows eruptions from the Sun's surface caused by particularly energetic fusion reactions.

Sections 7.7–7.11 Summary

- The mass defect of a nucleus is the difference between the total mass of the separate nucleons and the mass of the nucleus.
- Einstein's mass–energy equivalence relation: $E = mc^2$
- The binding energy of a nucleus is the energy needed to separate completely all its constituent nucleons.
- The binding energy per nucleon is a measure of the stability of a nucleus: a high binding energy per nucleon means the nucleus is stable.
- Nuclear fission is the splitting of a heavy nucleus into two smaller, lighter nuclei of approximately equal mass.
- Nuclear fusion is the joining together of light nuclei to form a larger, heavier nucleus.

Sections 7.7–7.11 Questions

1 Calculate the mass defect, the binding energy of the nucleus, and the binding energy per nucleon for the following nuclei:
  (a) hydrogen-3, ^3_1H; nuclear mass 3.01605 u,
  (b) zirconium-97, $^{97}_{40}\text{Zr}$; nuclear mass 97.09801 u,
  (c) radon-222, $^{222}_{86}\text{Rn}$; nuclear mass 222.01754 u.
  (proton mass = 1.00728 u; neutron mass = 1.00867 u)

2 One possible reaction taking place in the core of a reactor is

$$^{235}_{92}\text{U} + ^1_0\text{n} \rightarrow ^{95}_{42}\text{Mo} + ^{139}_{57}\text{La} + 2^1_0\text{n} + 7^{\ 0}_{-1}\text{e}$$

For this reaction, calculate:
  (a) the mass on each side of the equation,
  (b) the change in mass after fission has taken place,
  (c) the energy released per fission of uranium-235,
  (d) the energy available from the complete fission of 1.00 g of uranium-235,
  (e) the mass of uranium-235 used by a 500 MW nuclear power station in one hour, assuming 30% efficiency.
  (masses: $^{235}_{92}\text{U}$, 235.123 u; $^{95}_{42}\text{Mo}$, 94.945 u; $^{139}_{57}\text{La}$, 138.955 u; proton, 1.007 u; neutron, 1.009 u. 0.235 kg of uranium-235 contains 6.0×10^{23} atoms.)

3 Two fusion reactions which take place in the Sun are described below.
  (a) A hydrogen-2 (deuterium) nucleus absorbs a proton to form a helium-3 nucleus.
  (b) Two helium-3 nuclei fuse to form a helium-4 nucleus plus two free protons.
  For each reaction, write down the appropriate nuclear equation and calculate the energy released.
  (masses: ^2_1H, 2.01410 u; ^3_2He, 3.01605 u; ^4_2He, 4.00260 u; ^1_1p, 1.00728 u; ^1_0n, 1.00867u)

7.12 Probing matter

> At the end of Section 7.12 you should be able to:
> - infer from the results of the α-particle scattering experiment the existence and small size of the nucleus
> - select and use Coulomb's law to determine the force of repulsion, and Newton's law of gravitation to determine the force of attraction, between two protons at nuclear separations and hence understand the need for a short-range, attractive force between nucleons
> - estimate the density of nuclear matter.

Figure 7.18 Ion microscope photograph of iridium

Figure 7.18 shows a photograph taken with an ion microscope, a device which makes use of the de Broglie wavelength of gas ions (see page 183 of the AS book). It shows a sample of iridium at a magnification of about five million. The positions of individual iridium atoms can be seen.

Photographs like this reinforce the idea that all matter is made of very small particles we call atoms. Advances in science at the end of the nineteenth and the beginning of the twentieth century led most physicists to believe that atoms themselves are made from even smaller particles, some of which have positive or negative charges. Unfortunately even the most powerful microscopes cannot show us the internal structure of the atom. Many theories were put forward about the structure of the atom, but it was a series of experiments carried out by Ernest Rutherford and his colleagues around 1910 that led to the birth of the model we now know as the **nuclear atom**.

Probing matter using α-particles

In 1911, Rutherford and two of his associates, Geiger and Marsden, fired a beam of α-particles at a very thin piece of gold foil. A zinc sulphide detector was moved around the foil to see the directions in which the α-particles travelled after striking the foil (Figure 7.19).

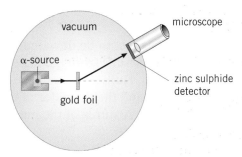

Figure 7.19 α-scattering experiment

They discovered that:

■ the vast majority of the α-particles passed through the foil with very little or no deviation from their original path
■ a small number of particles were deviated through an angle of more than about 10°
■ an extremely small number of particles (one in ten thousand) were deflected through an angle greater than 90°.

From these observations, the following conclusions could be drawn.

■ The majority of the mass of an atom is concentrated in a very small volume at the centre of the atom. Most α-particles would therefore pass through the foil undeviated.
■ The centre (or nucleus) of an atom is positively charged. α-particles, which are also positively charged, passing close to the nucleus will experience a repulsive force causing them to deviate.
■ Only α-particles that pass very close to the nucleus, almost striking it head-on, will experience large enough repulsive forces to cause them to deviate through angles greater than 90°. The fact that so few particles did so confirms that the nucleus is very small, and that most of the atom is empty space.
■ Because atoms are neutral, the atom must contain negative particles. These travel around, or orbit, the nucleus.

Figure 7.20 shows some of the possible trajectories of the α-particles. Using the nuclear model of the atom and equations to describe the force between charged particles, Rutherford calculated the fraction of α-particles that he

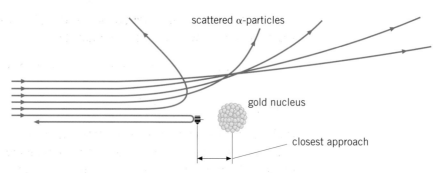

scattered α-particles

gold nucleus

closest approach

Figure 7.20

would expect to be deviated through various angles. The calculations agreed with the results from the experiment. This confirmed the nuclear model of the atom. Rutherford calculated that the diameter of the nucleus is about 10^{-15} m, and the diameter of the whole atom about 10^{-10} m. Figure 7.21 shows the features of the nuclear model of a nitrogen atom.

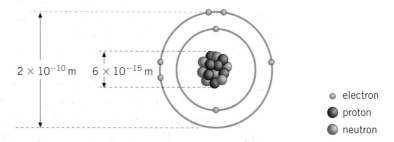

2×10^{-10} m 6×10^{-15} m

● electron
● proton
● neutron

Figure 7.21 The diameter of a nitrogen atom is more than 30 000 times bigger than the diameter of its nucleus

Some years later the α-particle scattering experiment was repeated using α-particles with higher energies. Some discrepancies between the experimental results and Rutherford's scattering formula were observed. These seemed to be occurring because the high-energy α-particles were passing very close to the nucleus, and were experiencing not only the repulsive electrostatic force but also a strong attractive force which appears to act over only a very short range. This became known as the **strong nuclear force**. This is the force that holds the nucleus together.

Could gravitation be responsible for this force? After all, the nucleons are so close together in the nucleus that the $1/r^2$ factor in the inverse square law becomes very important. Let's try a simple order-of-magnitude calculation to compare the force that could be provided by gravitation with the electrostatic repulsive force between protons in a nucleus. We will take the case of a helium nucleus, with its two protons, and assume that the distance between the protons is about 10^{-15} m, Rutherford's estimate of the diameter of the nucleus. The gravitational attractive force F_g between the protons, each of mass 1.7×10^{-27} kg, is

$$F_g = Gm^2/r^2 = 6.7 \times 10^{-11} \times (1.7 \times 10^{-27})^2/(1 \times 10^{-15})^2$$

leading to $F_g = 10^{-34}$ N

if we work in orders of magnitude only.

The corresponding calculation for the electrostatic repulsive force F_e, which also follows an inverse square law, is

$$F_e = (1/4\pi\epsilon_0)q^2/r^2 = (1/4\pi \times 8.9 \times 10^{-12}) \times (1.6 \times 10^{-19})^2/(1 \times 10^{-15})^2$$

leading to $\qquad F_e = 10^2\,\text{N}$

Perhaps surprisingly, this calculation shows that the magnitude of the gravitational attractive force is entirely negligible when compared with the magnitude of the electrostatic repulsive force, which is trying to pull the nucleus apart. This is our justification for saying that there must be another force, the strong nuclear force, acting in opposition to the electrostatic repulsive force. The characteristic of this force is that it must be stronger than the electrostatic force, or else nuclei would not hold together. But it is also a short-range force; for example, its effects are not felt by the electrons in the atom as a whole. In the simple model of electrons circulating in orbits around the nucleus, it is the electrostatic attractive force between the negatively-charged electrons and the positively-charged nucleus that provides the centripetal force. Outside the nucleus, the strong nuclear force is not felt at all, so as a rough estimate we can say that it has a range of about $10^{-15}\,\text{m}$.

There is another interesting consequence of the strong nuclear force. It holds the nucleons together in such a compact body that the volume of the nucleus is very small indeed. This means that the density of nuclear matter is extraordinarily high, compared with that of macroscopic materials. Taking the helium nucleus as an example, it contains four nucleons so that its mass is about $4 \times 1.7 \times 10^{-27} = 6.8 \times 10^{-27}\,\text{kg}$. Again assuming that the nucleus is spherical and using Rutherford's estimate of the diameter of the nucleus as about $10^{-15}\,\text{m}$, the volume of the nucleus is about $5 \times 10^{-46}\,\text{m}^3$. This means that the density of the helium nucleus is roughly $10^{-26}/10^{-45} = 10^{19}\,\text{kg m}^{-3}$. Compare this with the density of lead, which is $1.1 \times 10^4\,\text{kg m}^{-3}$!

Probing matter using electrons

Electrons are not affected by the strong nuclear force. It was suggested that they might therefore be a more effective tool with which to investigate the structure of the atom. As we saw in Chapter 7 of the AS book, moving electrons have a wavelike property, and can be diffracted.

If a beam of electrons is directed at a sample of powdered crystal and the electron wavelength is comparable with the interatomic spacing in the crystal, the electron waves are scattered from planes of atoms in the tiny crystals, creating a diffraction pattern (Figure 7.22). The fact that a diffraction pattern is obtained confirms the regular arrangement of the atoms in a crystalline solid. Measurements of the angles at which strong scattering is obtained can be used to calculate the distances between planes of atoms.

If the energy of the electron beam is increased, the wavelength decreases. Eventually the electron wavelength may be of the same order of magnitude as the diameter of the nucleus. Probing the nucleus with high-energy electrons rather than α-particles gives a further insight into the dimensions of the nucleus, and also gives information about the distribution of charge in the nucleus itself.

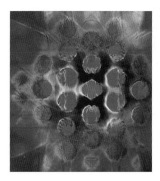

Figure 7.22 Electron diffraction pattern of a sample of pure titanium

Section 7.12 Summary

- The Rutherford α-particle experiment confirmed the nuclear model of the atom: the atom consists of a small, heavy, positively-charged nucleus, surrounded by negatively-charged electrons in orbit about the nucleus.
- The diameter of the nucleus is about 10^{-15} m; the diameter of the atom is about 10^{-10} m. The density of nuclear matter is about $10^{19}\,\mathrm{kg\,m^{-3}}$. Nucleons in the nucleus are held together by the strong nuclear force.
- Electron diffraction gives evidence for the regular arrangement of atoms in crystals, and allows the measurement of the distance between planes of atoms in solids.

Section 7.12 Questions

1 It is found that radii r of nuclei depend on their nucleon numbers A according to the relation $r = r_0 A^{1/3}$, where $r_0 \approx 1.2 \times 10^{-15}\,\mathrm{m}$. Find the ratio of the radius of a nucleus of ^{238}U to the radius of a nucleus of ^{1}H.

2 Explain the evidence for the existence of the strong nuclear force.

7.13 Fundamental particles

At the end of Section 7.13 you should be able to:
- understand what is meant by a fundamental particle
- explain that since protons and neutrons contain charged constituents called quarks they are not fundamental particles
- describe a simple quark model of hadrons in terms of up, down and strange quarks and their respective antiquarks, taking into account their charge, baryon number and strangeness
- describe how the quark model may be extended to include the properties of charm, topness and bottomness
- describe the properties of neutrons and protons in terms of a simple quark model
- describe how there is a weak interaction between quarks, and that this is responsible for β-decay
- understand that there are two types of β-decay
- describe the two types of β-decay in terms of a simple quark model
- know that (electron) neutrinos and (electron) antineutrinos are produced during β^+ and β^- decays, respectively
- know that a β^- particle is an electron and a β^+ particle is a positron
- know that electrons and neutrinos are members of a group of particles known as leptons.

By the mid-1930s, the nuclear model of the atom was well established. It was known that the nucleus consisted of protons and neutrons, while outside the nucleus were electrons. These were considered to be the sub-nuclear particles which were three of the basic building blocks of the universe. Besides these three, three others were known: the positron (similar to the electron but with a positive charge of $+e$), the neutrino (a particle with zero charge and with a mass, when at rest, of zero), and the γ-particle (a photon). This gave a total of six sub-nuclear particles. At this time the structure of matter seemed straightforward.

However, since then many more sub-nuclear particles were discovered. It became clear that not all could possibly be fundamental (or elementary, the alternative description). The study of these particles, their classification, interactions and properties, and whether they were indeed fundamental, is the subject of **fundamental particle physics**.

Matter and antimatter

The positron has just been described as a positive electron. It has the same mass as the electron but a charge of $+e$. It is called the *antiparticle* of the electron. When it was discovered, in 1932, it was predicted that other particles, such as the proton, should have corresponding antiparticles. It was 1955 before the antiproton, a particle with the proton mass but a charge of $-e$, was found. In particle physics equations the proton is given the symbol p, and the antiproton $\bar{\text{p}}$ ('p-bar'); in general, an antiparticle is distinguished by a bar over the particle symbol. It might seem a pleasing symmetry if all particles had corresponding antiparticles, but this is not the case; for example, the photon does not have an antiphoton.

Generally, antiparticles need a great deal of energy for their production, and this is why physicists look for them as a result of nuclear reactions. Their lifetime is very short in the presence of matter. An isolated positron is stable, but when it encounters an electron, they annihilate one another, the energy produced being emitted in the form of two gamma-photons.

Conservation laws in particle reactions

We have already met the laws of conservation of energy, momentum and electric charge. These are fundamental laws of physics and also apply in reactions between particles. But there are some reactions which, if they ever took place, would obey these conservation laws, but which in fact never occur. A fourth conservation law, the **law of conservation of baryon number**, has to be invoked.

For particles like the proton and neutron, the baryon number is numerically the same as nucleon number (mass number). We have applied the conservation of nucleon number in reactions between nuclei and in nuclear decay (see Section 7.2). In particle physics the nucleons have baryon number $B = +1$, while their antiparticles have baryon number $B = -1$.

A negative baryon number does not mean that antiparticles have negative mass; it just signals their antimatter status.

In order to explain the properties of a particular type of particle which had been given the descriptor **strange**, it also proved necessary to introduce a new quantum number, **strangeness**. This has its own conservation law, the **law of conservation of strangeness**.

Classification of fundamental particles

The starting point for classifying sub-nuclear particles is a principle relating to how electrons in many-electron atoms (that is, all atoms except hydrogen) occupy quantum states. This is the **Pauli exclusion principle**, which says that no two electrons in an atom can occupy the same quantum state.

It is found that all observed sub-nuclear particles fall into two classes. One class is the **fermions**, which obey the Pauli exclusion principle. (The electron is, by definition, a fermion.) The other class is the **bosons**, which do not obey the exclusion principle.

Fermions are themselves divided into two sorts, **leptons** and **hadrons**. Hadrons interact through the strong nuclear force (see Section 7.10), whereas leptons interact through the weak nuclear force (as well as the much weaker gravitational force), and the electromagnetic force if they are charged. In the 1960s, when this classification was proposed, there were only four known leptons, of which two were the electron and the neutrino. Since 1960 a few more leptons have been added, mainly because it was realised that the neutrino discovered in the 1930s was a specific sort of neutrino, the electron neutrino, and that there are a number of different types of neutrino, all of which are leptons.

Leptons seem to have no measurable size and do not break down into smaller particles. For this reason, they are considered to be fundamental particles. On the other hand, there are very many hadrons, so many that it is most unlikely that they are fundamental.

Remember that protons and neutrons are held together in the nucleus by the strong nuclear force? Hence these two must be hadrons. In 1963 Murray Gell-Mann proposed that hadrons could not be classed as fundamental particles. He suggested that all hadrons were composed of combinations of three fundamental entities called **quarks**.

Quarks

In a simple model, we consider there to be three quarks, with their corresponding antiquarks. The quarks are named **up**, **down** and **strange**, with symbols u, d and s respectively. The antiquarks are $\bar{u}$, $\bar{d}$ and $\bar{s}$. They have a fractional charge of either one-third or two-thirds of the electron charge. Note that this means that what we had previously considered to be the fundamental unit of charge, the elementary charge, has actually been divided. Quarks have a baryon number of $+\frac{1}{3}$, and antiquarks $-\frac{1}{3}$.

Shortly after Gell-Mann's proposals were published, it became obvious that the three-quark model was insufficient to explain the properties of some recently-discovered hadrons; the model would have to be extended to include three more quarks, each with another quantum number and conservation law. The names of these quarks are the **charmed** (c), **bottom** (b) and **top** (t) quark. Again, each quark has its antiquark.

It would be pleasing to detect a free quark experimentally but, perhaps because they are so tightly bound together, this has proved impossible. We must be satisfied with indirect evidence for their existence.

Table 7.6 summarises the properties of quarks and antiquarks.

Table 7.6 The properties of quarks and antiquarks

name (symbol)	charge	baryon number	strangeness	charm	bottomness	topness
Quarks						
up (u)	$+2e/3$	1/3	0	0	0	0
down (d)	$-e/3$	1/3	0	0	0	0
strange (s)	$-e/3$	1/3	-1	0	0	0
charmed (c)	$+2e/3$	1/3	0	$+1$	0	0
bottom (b)	$-e/3$	1/3	0	0	-1	0
top (t)	$+2e/3$	1/3	0	0	0	$+1$
Antiquarks						
up ($\bar{u}$)	$-2e/3$	$-1/3$	0	0	0	0
down ($\bar{d}$)	$+e/3$	$-1/3$	0	0	0	0
strange ($\bar{s}$)	$+e/3$	$-1/3$	$+1$	0	0	0
charmed ($\bar{c}$)	$-2e/3$	$-1/3$	0	-1	0	0
bottom ($\bar{b}$)	$+e/3$	$-1/3$	0	0	$+1$	0
top ($\bar{t}$)	$-2e/3$	$-1/3$	0	0	0	-1

Hadrons are divided into two sub-groups: baryons, which have baryon number +1 (or −1 for their antiparticles), and mesons, with baryon number 0. Baryons need three quarks to describe them. For example, a proton is p = uud and a neutron is n = udd. The antiproton is $\bar{p} = \overline{uud}$ and the antineutron is $\bar{n} = \overline{uud}$. The official definition of the baryon number of a particle or antiparticle is

(number of quarks in the particle – number of antiquarks in it)/3.

Gell-Mann received the 1969 Nobel Prize for his work on the quark explanation of fundamental particles.

Example

Confirm that the proton, when considered on the quark model, has the expected baryon number and charge.

The proton is p = uud.

Consider the baryon number first. Both up and down quarks have baryon number +1/3, so the baryon number of the proton, containing two up quarks and one down quark and no antiquarks, is **+1**, as expected.

Now for the charge. The charge on an up quark is $+2e/3$, and that on the down quark is $-e/3$, so the resultant charge is $(2e/3 + 2e/3 - e/3) = $ **+e**, again as expected.

Now it's your turn

1 Confirm that the neutron (n = udd), when considered on the quark model, has the expected baryon number and charge.

2 Use the quark model to deduce the baryon number and charge of the antiproton ($\bar{p} = \overline{uud}$).

Beta decay in terms of quarks

In Section 7.2 the process of β-emission was described as the emission of an electron (a β-particle) from a radioactive nucleus. To be precise, we should really call it β^- emission, to emphasise that it is a negatively-charged electron that is emitted. To complete the picture, we should state that positron emission, or β^+ emission, may also be obtained.

Quarks are fundamental particles and thus interact by the weak nuclear interaction. The weak interaction can cause the conversion of a neutron in a radioactive nucleus into a proton, while emitting an electron and a particle called an electron antineutrino.

$$^1_0 n \rightarrow\ ^1_1 p\ +\ ^{\ 0}_{-1} e\ +\ \bar{\nu}_e$$

(In Section 7.2, where we introduced this idea of neutron conversion to a proton and electron, we cheated a little by ignoring the antineutrino and merely adding 'energy' to the right-hand side of the equation. In fact, we need another particle to carry off energy and momentum and satisfy the conservation laws. This particle is the antineutrino.)

At the quark level, the process of neutron (n = udd) decay is described by the decay of one of the neutron's down quarks to an up quark. An intermediate particle, the W^- boson, is emitted; this decays into an electron (the β^- particle) and an electron antineutrino.

Positron emission is, at first sight, simply the positive analogue to β^- emission. One of the up quarks in the proton goes to a down quark, producing a neutron, a positron (the β^+ particle) and a neutrino. Note that in positron emission it is a neutrino that is emitted with the positron, not the antineutrino of electron emission. We describe the process by the equation

$$_1^1\text{p} \rightarrow {}_0^1\text{n} + {}_1^0\text{e} + \nu_\text{e}$$

But because the mass of the neutron (plus the electron) is greater than the mass of the proton, the reaction cannot precede spontaneously in the way that β^- emission can; it requires the input of energy on the left-hand side of the equation. This can occur only when the binding energy of the mother nucleus is less than that of the daughter nucleus. Rather few mother nuclei fit this requirement.

Section 7.13 Summary

- A fundamental particle is a sub-nuclear particle that does not contain any constituent particles.
- Many sub-nuclear particles have a corresponding antiparticle.
- The Pauli exclusion principle states that, in a many-electron atom, no two electrons can occupy the same quantum state; fermions are sub-nuclear particles that obey the Pauli principle applied to nuclear quantum states.
- Fermions are classified into leptons, which interact through the weak nuclear force, have no sub-structure and are truly fundamental particles, and hadrons, which interact through the strong nuclear force, have a sub-structure and are not fundamental particles.
- Protons and neutrons are hadrons, since they are held together in the nucleus by the strong nuclear force.
- Hadrons are composed of fundamental particles called quarks, which have non-integral electric charge.

Section 7.13 Questions

1 **(a) (i)** What is meant by a *fundamental particle?*
   (ii) From the following list of subatomic particles, state the particle(s) that are *not* fundamental: neutron, neutrino, electron, proton
 (b) Leptons and hadrons are names given to two groups of sub-atomic particles.
   (i) State a difference between leptons and hadrons.
   (ii) Hadrons are further subdivided into two classes of particle. Name the classes of particle concerned.

   (iii) Particles that are not fundamental are classified in terms of their quark structure. State the general difference, in terms of quark structure, between the two classes of particle you have named in **(b)(ii)**. For two of the particles in the list in **(a)(ii)** that have a quark structure, give details of these structures.

Exam-style Questions

1 Cobalt-60 is often used as a radioactive source in medical physics. It has a half-life of 5.25 years. Calculate how long after a new sample is delivered the activity will have decreased to one-eighth of its value when delivered.

2 Iodine-131 is a beta-emitter with half-life of 194 hours. It decays to xenon (Xe). The decay is represented by

$$^{131}_{53}I \rightarrow ^{P}_{Q}Xe + \beta$$

(a) Obtain the values of P and Q in the decay equation.

(b) A sample of iodine-131 of mass 0.20 g is prepared in a reactor.
   (i) Given that 0.131 kg of iodine-131 contains 6.02×10^{23} atoms, calculate the number of iodine-131 atoms contained in the sample at the time of preparation.
   (ii) Calculate the mass of iodine-131 which will have decayed 48 hours after preparation.

3 The half-life of a radioactive isotope of sodium used in medicine is 15 hours.

(a) Determine the decay constant for this nuclide.

(b) A small volume of a solution containing this nuclide has an activity of 1.2×10^4 disintegrations per minute when it is injected into the bloodstream of a patient. After 30 hours, the activity of $1.0\,cm^3$ of blood taken from the patient is 0.50 disintegrations per minute. Estimate the volume of blood in the patient. Assume that the solution is uniformly diluted in the blood, that it is not taken up by the body tissues, and that there is no loss by excretion.

4 A fusion reaction that takes place in the Sun is represented by the equation

$$^{2}_{1}H + ^{2}_{1}H \rightarrow ^{3}_{2}He + ^{1}_{0}n + energy$$

(a) Calculate the energy released in this reaction. (masses: $^{2}_{1}H$, 2.0141 u; $^{3}_{2}He$, 3.0160 u; $^{1}_{0}n$, 1.0087 u)

(b) Suggest why it is difficult to achieve controlled reactions of this type.

5 A stationary radium nucleus ($^{224}_{88}Ra$) of mass 224 u spontaneously emits an α-particle. The α-particle is emitted with an energy of 9.2×10^{-13} J, and the reaction gives rise to a nucleus of radon (Rn).

(a) Write down a nuclear equation to represent α-decay of the radium nucleus.

(b) Show that the speed with which the α-particle is ejected from the radium nucleus is 1.7×10^7 m s^{-1}.

(c) Calculate the speed of the radon nucleus on emission of the α-particle. Explain how the principle of conservation of momentum is applied in your calculation.

6 When an α-particle travels through air, it loses energy by ionisation of air molecules. For every air molecule ionised, approximately 5.6×10^{-18} J of energy is lost by the α-particle.

(a) Suggest a typical value for the range of an α-particle in air. Hence estimate the number of air molecules ionised per millimetre of the path of the α-particle, given that the α-particle has initial energy 9.2×10^{-13} J.

(b) It has been discovered that the number of ionisations per unit length of the path of an α-particle suddenly increases just before the α-particle stops. State, with a reason, the effect that this observation will have on your estimate.

7 (a) Consider a neutral atom of beryllium ($^{9}_{4}Be$). State how many of each type of particle, leptons and baryons, a neutral atom of beryllium contains. Explain your answer.

(b) The following equation represents the reaction between a particle X and a neutron.

$$X + n \rightarrow p + e^-$$

In this equation, lepton number is conserved in the same way as charge is conserved. That is, the sum of the lepton numbers on the left-hand side of the equation must equal the sum of the lepton numbers on the right-hand side. By consideration of the conservation of charge and the conservation of lepton number, identify particle X.

8. Medical imaging

Have you ever broken a bone? If you have, then you will probably have had an X-ray image of the broken bones taken at a hospital. From the image a doctor would then know how best to set the bones in place to heal properly. Dentists make use of X-ray images during treatment and to check for infections. An X-ray image is an example of one of the many imaging techniques now available to medical science.

8.1 Methods of medical diagnosis

At the end of Section 8.1 you should be able to:
- describe the need for non-invasive techniques in diagnosis.

Over a century ago, medical diagnosis was achieved using two techniques. The first technique was to observe the patient externally for signs of fever, changed pulse-rate, changed breathing, vomiting, skin condition, and so on. Diagnosis on this basis was part science and part art. The alternative was to carry out investigative surgery. Surgery always involves a risk and, in the past, many patients died either from the trauma of surgery or from resulting infection.

Diagnostic techniques have now been developed that enable externally placed equipment to obtain information from under the skin. X-ray diagnosis was developed throughout the twentieth century. X-ray exposure is not without some risk, however small, and during the second half of the last century other techniques evolved. The use of ultrasound and magnetic resonance imaging are now commonplace.

Using these imaging techniques, it is possible to obtain detailed information about internal body structures without surgery, and consequently the techniques are referred to as being *non-invasive*. They are designed to cause as little risk as possible to the patient. It should be appreciated, however, that there is some risk with any procedure. The risks involved in imaging diagnoses are, in general, very small. These very small risks have to be weighed against the benefits of their use and the real finite risk that accompanies any disease.

8.2 X-rays

At the end of Section 8.2 you should be able to:
- describe the nature of X-rays
- describe in simple terms how X-rays are produced
- describe how X-rays interact with matter (limited to photoelectric effect, Compton effect and pair production)
- define **intensity** as power per unit cross-sectional area
- select and use the equation $I = I_0 e^{-\mu x}$ to show how the intensity I of a collimated X-ray beam varies with thickness x of medium
- describe the use of X-rays in imaging internal body structures including the use of image intensifiers and of contrast media
- explain how soft tissues like the intestines can be imaged using a barium meal
- describe the use of a computed axial topography (CAT) scanner
- describe the advantages of a CAT scan compared with an X-ray image.

The production of X-rays

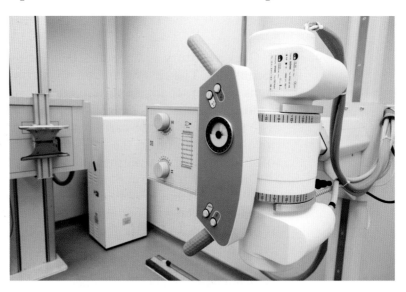

Figure 8.1 Modern X-ray machine

Whenever a charged particle is accelerated, electromagnetic radiation is emitted. This radiation is known as Bremmstrahlung radiation or 'braking (slowing down)' radiation. The frequency of the radiation depends on the magnitude of the acceleration. The larger the acceleration (or deceleration), the greater is the frequency of the emitted photon.

X-ray photons may be produced by the bombardment of metal targets with high-speed electrons. The electrons are first accelerated through a potential difference of several thousands of volts so that they have high energy and high speed. This acceleration is, however, not sufficient for X-ray radiation to be emitted. The high-speed electrons strike a metal target, which causes the electrons to change direction and to lose kinetic energy very rapidly. Large decelerations are involved that give rise to the emission of X-ray photons. These X-ray photons have wavelengths shorter than about 5×10^{-8} m. It should be remembered that not all of the energy of the electrons is emitted as X-ray photons. The majority is transferred to thermal energy in the target metal.

A simplified design of an X-ray tube is shown in Figure 8.2. A metal filament is heated using a low-voltage supply and electrons are emitted from the filament – this is called the thermionic effect. The electrons are then accelerated towards the target, constructed of metal with a high melting point. The electrons are accelerated through a large potential difference of the order of 20–90 kV. When the electrons are stopped in the metal target, X-ray photons (Bremmstrahlung radiation) are emitted and the X-ray beam passes out of the tube through a window that is transparent to X-rays. The tube is evacuated and is made of a material that is opaque to X-rays. This reduces background radiation around the tube. The anode is earthed and the cathode is maintained at a high negative potential, so that the equipment is at earth potential. Since the majority of the energy of the electrons is transferred to thermal energy in the anode, the anode is either cooled or is made to spin rapidly so that the thermal energy is dispersed over a larger area of the target.

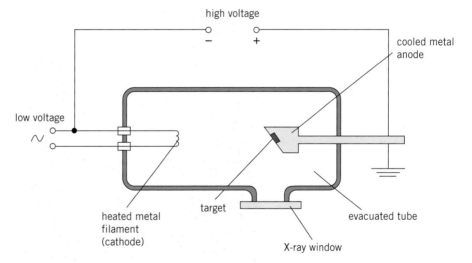

Figure 8.2 Design of an X-ray tube

Control of the X-ray beam

In order that the optimum X-ray image may be obtained, both the intensity and the hardness of the X-ray beam need to be controlled.

- The **intensity** is the wave power per unit cross-sectional area of the beam, and this affects the degree of blackening of the image.
- The **hardness** is the penetration of the X-ray beam, which determines the fraction of the intensity of the incident beam that can penetrate the part of the body being X-rayed. In general, the shorter the wavelength of the X-rays, the greater their penetration.

A typical X-ray spectrum showing the variation with wavelength of the intensity of an X-ray beam is shown in Figure 8.3.

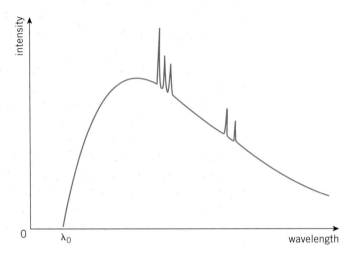

Figure 8.3 Typical X-ray spectrum

The spectrum has two distinct components. First, there is a continuous distribution of wavelengths with a sharp cut-off at the shortest wavelength, λ_0. Second, sharp peaks may be observed. These sharp peaks correspond to the emission line spectrum of the target metal and are, therefore, a characteristic of the target.

The continuous distribution comes about because the electrons, when incident on the metal target, will not all have the same deceleration but will, instead, have a wide range of values. Since the wavelength of the emitted radiation is dependent on the deceleration, there will be a distribution of wavelengths. The cut-off wavelength corresponds to an electron that is stopped in one collision in the target so that all of its kinetic energy is given up as one X-ray photon.

The kinetic energy E_K of an electron is equal to the energy gained by the electron when it is accelerated from the cathode to the anode.

$$E_K = eV$$

where e is the charge on the electron and V is the accelerating potential difference.

The energy E of a photon of wavelength λ is given by

$$E = \frac{hc}{\lambda}$$

where h is the Planck constant (see page 178 of the AS book). Thus, at the cut-off wavelength,

$$eV = \frac{hc}{\lambda_0}$$

and so

$$\lambda_0 = \frac{hc}{eV}$$

The accelerating potential V thus determines the cut-off wavelength λ_0. The larger the potential difference, the shorter the wavelength. The hardness of the X-ray beam is therefore controlled by variation of the accelerating potential difference between the cathode and the anode.

The continuous distribution of wavelengths implies that there will be X-ray photons of long wavelengths that would not penetrate the person being investigated and so would not contribute towards the X-ray image. Such long-wavelength photons would add to the radiation dose received by the person without serving any useful purpose. For this reason the X-ray beam emerging from the X-ray tube frequently passes through aluminium filters that absorb these long-wavelength photons.

The intensity of the X-ray beam depends on the number of photons emitted per unit time and hence the number of electrons hitting the metal target per unit time. Since the electrons are produced by thermionic emission, then increasing the heater current in the cathode will increase the rate of production of electrons and hence increase the intensity of the X-ray beam.

Example

The accelerating potential difference between the cathode and the anode of an X-ray tube is $30\,\text{kV}$. Given that the Planck constant is $6.6 \times 10^{-34}\,\text{J}\,\text{s}$, the charge on the electron is $1.6 \times 10^{-19}\,\text{C}$ and the speed of light in free space is $3.0 \times 10^{8}\,\text{m}\,\text{s}^{-1}$, calculate the minimum wavelength of photons in the X-ray beam.

For the minimum wavelength,
energy gained by electron = energy of photon

$$eV = \frac{hc}{\lambda_0}$$

$$1.6 \times 10^{-19} \times 30 \times 10^3 = \frac{(6.6 \times 10^{-34} \times 3.0 \times 10^8)}{\lambda_0}$$

$$\lambda_0 = \textbf{4.1} \times \textbf{10}^{-11}\,\textbf{m}$$

Now it's your turn

(a) State what is meant by the *hardness* of an X-ray beam.
(b) The Planck constant is $6.6 \times 10^{-34}\,\text{J}\,\text{s}$, the charge on the electron is $1.6 \times 10^{-19}\,\text{C}$ and the speed of light in free space is $3.0 \times 10^{8}\,\text{m}\,\text{s}^{-1}$. Calculate the minimum wavelength of photons produced in an X-ray tube for an accelerating potential difference of $45\,\text{kV}$.

Attenuation mechanisms

Whenever a beam of X-ray radiation passes through matter, the photons lose energy and the intensity of the beam decreases. The beam is said to be **attenuated**. This absorption of energy occurs as a result of photons interacting with atoms of the substance. The main processes, for X-ray energies used for medical purposes are

- the photoelectric effect
- the Compton effect
- pair production.

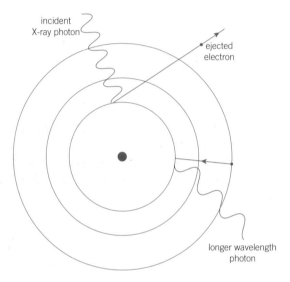

Figure 8.4 The photoelectric effect

The photoelectric effect

The photoelectric effect is illustrated in Figure 8.4.

The X-ray photon has an energy greater than the binding energy of an electron in an atom. During the interaction between the photon and the atom, the electron is ejected from the atom. The atom is ionised and the photon is annihilated. Another electron in the atom that is in a higher energy state than that which was occupied by the ejected electron now moves to fill the vacancy. A photon with a lower energy than the incident photon is now emitted.

Compton effect

This effect is sometimes referred to as Compton scattering. It is illustrated in Figure 8.5.

The photon energy is much greater than the binding energy of the electron in an atom. Essentially, the photon 'sees' the electrons as being free.

An electron is ejected from the atom, causing ionisation. The photon changes direction (in order to conserve momentum) and moves off with a lower energy (to conserve energy). The wavelength of the photon has increased. This lower energy photon may now lose more energy by the Compton effect or the photoelectric effect.

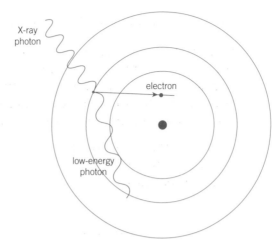

Figure 8.5 Compton effect

Pair production

For this effect to occur, the X-ray photons must have an energy in excess of about 1 MeV.

The X-ray photon interacts with the nucleus of an atom, giving rise to an electron and a positron. This is illustrated in Figure 8.6.

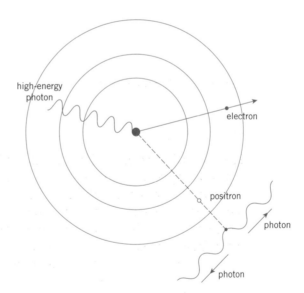

Figure 8.6 Pair production

The positron has a mass equal to that of the electron. The charge on the positron is equal to that of the electron but of opposite sign.

As the electron and the positron move off, they lose energy by ionisation. The positron will be annihilated by combination with an electron, producing two γ-ray photons, each of energy 0.51 MeV. These γ-ray photons will then lose energy by Compton scattering and subsequently by the photoelectric effect.

The attenuation of X-rays

When a beam of X-ray photons passes through a medium, absorption processes occur that reduce the intensity of the beam. The intensity of a parallel beam is reduced by the same fraction each time the beam passes through equal thicknesses of the medium. Consequently, the variation of the percentage of the intensity transmitted with thickness of absorber may be shown as in Figure 8.7.

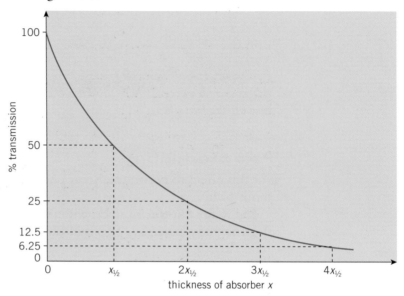

Figure 8.7 The percentage transmission of X-rays in a medium

It can be seen that the same thickness of medium is always required to reduce the transmitted beam intensity by 50%, no matter what starting point is chosen. This thickness of medium is called the **half-value thickness** (HVT) and is denoted by the symbol $x_{\frac{1}{2}}$.

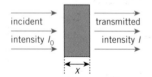

Figure 8.8 Absorption of X-rays in a medium

The decrease in transmitted intensity is an exponential decrease. Consider a parallel beam having an incident intensity I_0. The medium (the absorber) has thickness x and the transmitted intensity is I, as illustrated in Figure 8.8. The transmitted intensity is given by the expression

$$I = I_0\, e^{-\mu x} \quad \text{or} \quad I = I_0 \exp(-\mu x)$$

where μ is a constant that is dependent on the medium and on the energy of the X-ray photons, known as the **linear attenuation coefficient** or the **linear absorption coefficient** of the medium. The unit of μ is mm^{-1} or cm^{-1}.

For a thickness $x_{\frac{1}{2}}$ (the half-value thickness) of the medium, the intensity I will be equal to $\frac{1}{2}I_0$. Hence,

$$\tfrac{1}{2}I_0 = I_0 \exp(-\mu x_{\frac{1}{2}})$$

which gives

$$\ln 2 = \mu x_{\frac{1}{2}}$$

Approximate values of the linear attenuation (absorption) coefficient μ for some substances are given in Table 8.1.

Table 8.1 Some approximate values of linear attenuation (absorption) coefficient

substance	μ/cm^{-1}
copper	7
water	0.3
bone	3
fat	0.9

Example

The linear absorption coefficient of copper is $0.693\,\mathrm{mm}^{-1}$. Calculate:
(a) the thickness of copper required to reduce the incident intensity by 50% (the half-value thickness),
(b) the fraction of the incident intensity of a parallel beam that is transmitted through a copper plate of thickness 1.2 cm.

(a) $I = I_0 e^{-\mu x}$ or $I = I_0 \exp(-\mu x_{\frac{1}{2}})$

$I/I_0 = 0.50 = \exp(-0.693 x_{\frac{1}{2}})$

$\ln 0.50 = -0.693 x_{\frac{1}{2}}$

$x_{\frac{1}{2}} = \mathbf{1.0\,mm}$

(b) $I/I_0 = \exp(-0.693 \times 12)$
$I/I_0 = \mathbf{2.4 \times 10^{-4}}$

Now it's your turn

1 For one particular energy of X-ray photons, water has a linear attenuation (absorption) coefficient of $0.29\,\mathrm{cm}^{-1}$. Calculate the depth of water required to reduce the intensity of a parallel beam of these X-rays to 1.0×10^{-3} of its incident intensity.

2 The linear attenuation (absorption) coefficients of bone and of the soft tissues surrounding the bone are $2.9\,\mathrm{cm}^{-1}$ and $0.95\,\mathrm{cm}^{-1}$ respectively. A parallel beam of X-rays is incident, separately, on a bone of thickness 3.0 cm and on soft tissue of thickness 5.0 cm. Calculate the ratio:

$$\frac{\text{intensity transmitted through bone}}{\text{intensity transmitted through soft tissue}}$$

The X-ray image

An X-ray image is shown in Figure 8.9. This is not really an image in the sense of the real image produced by a lens. Rather, the image is like a shadow, as illustrated in Figure 8.10.

MEDICAL IMAGING

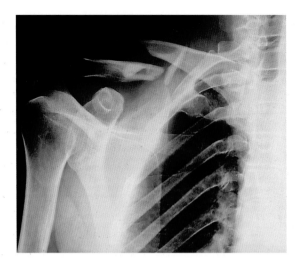

Figure 8.9 X-ray image showing a fractured collar bone

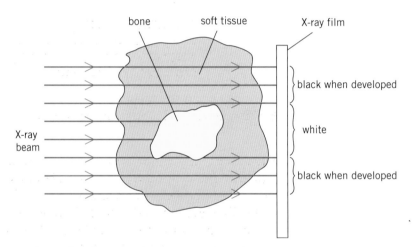

Figure 8.10 How an X-ray image is produced

The X-ray beam is incident on the body part of the patient. The X-ray beam can penetrate soft tissues (skin, fat, muscle, etc.) with little loss of intensity and so photographic film, after development, will show a dark area corresponding to these soft tissues. Bone, however, causes a greater attenuation (reduces the intensity by a greater extent) than the soft tissues and therefore the photographic film will be lighter in colour in areas corresponding to the positions of bones. What is produced on the film is a two-dimensional shadow of the bone and the surrounding tissues.

The quality of the shadow image produced depends on its sharpness and contrast. **Sharpness** is related to the ease with which the edges of structures can be determined. A shadow image where the bones and other organs are clearly outlined is said to be a 'sharp image'. Although an image may be sharp, it may still not be clearly visible because there is little difference in the degree of blackening between, say, the bone and the surrounding tissue. An X-ray image having a wide range of degrees of blackening is said to have good **contrast**.

Good contrast is achieved when neighbouring body organs and tissues absorb the X-ray photons to very different extents. This is usually the case for

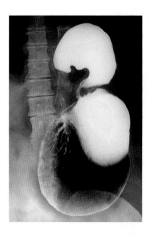

Figure 8.11 X-ray of stomach after a barium meal

bone and muscle. This is, however, not the case where, for example, the stomach is to be investigated. The patient is then asked to swallow a solution of barium sulphate – a 'barium meal' (Figure 8.11). The barium is a good absorber of X-ray photons. As a result, when the barium sulphate solution coats the inside of the stomach, the outline of the stomach will be shown up clearly on the image. Similarly, blood vessels can be made visible by injecting a radio-opaque dye into the bloodstream.

Contrast also depends on other factors, such as exposure time. Contrast may be improved by using intensifying screens in conjunction with a photographic film.

Intensifying screens not only improve contrast but also reduce exposure time, thus reducing the radiation dose received by the patient. The X-ray plate consists of a double-sided photographic film that is sandwiched between two intensifying screens, as shown in Figure 8.12.

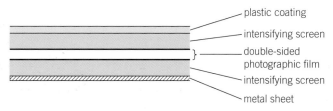

Figure 8.12 Photographic film with intensifying screens

The intensifying screens contain a fluorescent material such as zinc sulphide crystals. When exposed to X-ray radiation, the X-ray photons are absorbed by electrons in the crystals and, when the electrons de-excite, photons of visible light are emitted. These photons then affect the photographic film which is more sensitive to visible light than to X-ray radiation. The exposure time may be reduced by a factor of about 100.

Computed axial tomography (CAT or CT scanning)

The image produced on X-ray film as outlined earlier is a 'shadow' or 'flat' image. There is little, if any, indication of depth. That is, the position of an organ within the body is not apparent. Also, soft tissues lying behind structures that are very dense cannot be detected. Tomography is a technique whereby a three-dimensional image or 'slice' through the body may be obtained. The image is produced by **computed axial tomography** or **computed tomography** using what is known as a CAT or CT scanner (Figure 8.13). An example of such an image is shown in Figure 8.14.

In this technique, a series of X-ray images is obtained. Each image is taken through the section or slice of the body from a different angle, as illustrated in Figure 8.15.

Data for each individual X-ray image and angle of viewing is fed into a high-power computer. The computer enables the image of each slice to be combined so that a complete three-dimensional image of the whole object is obtained which can then be viewed from any angle.

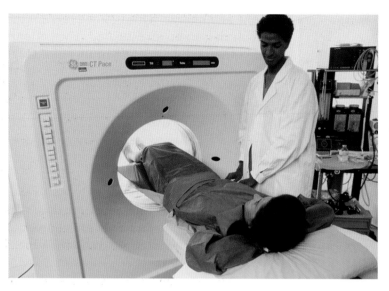

Figure 8.13 Radiologist moving a patient into a CAT scanner

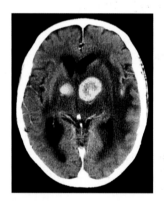

Figure 8.14 CAT scan through the head of a patient with a cerebral lymphoma

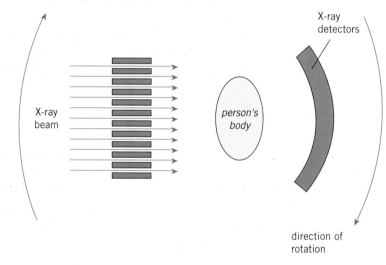

Figure 8.15 Arrangement for a CAT scan

All the data for all the sections is stored in the computer memory so that a three-dimensional image of the whole object is formed. This enables sections of the image to be viewed from many different angles. The computer also enables the brightness and contrast of the image to be varied so that the optimum image may be obtained.

Example

Compare the image produced during an X-ray investigation and that produced in CAT scanning.

X-ray image is a two dimensional projection onto a flat screen of a three-dimensional object. A CAT scan is a three-dimensional image. The computer in which the data for the image is stored enables different sections at different angles to be viewed.

Now it's your turn

The principles of CAT scanning have been understood for some time. However, scanners could not be developed until large powerful computers were made available. By reference to the image produced in a CAT scan, suggest why such a computer is necessary.

Sections 8.1–8.2 Summary

- Remote sensing enables investigations to be made where there is no contact with the object under investigation.
- X-rays are produced when high-speed electrons are stopped by a metal target.
- An X-ray image is a 'shadow' of structures in which the X-ray beam is attenuated.
- The attenuation in the intensity of an X-ray beam is given by:
  $I = I_0 \, e^{-\mu x}$
- Computed axial tomography (CAT scanning) enables an image of a section through the body to be obtained by combining many X-ray images, each one taken from a different angle.

Sections 8.1–8.2 Questions

1 (a) State what is meant by the *sharpness* of an X-ray image.
   (b) Explain how the sharpness of an X-ray image may be improved.

2 (a) Outline how X-ray images are used to build up the image produced in a CAT scan.
   (b) Explain why the radiation dose received during a CAT scan is greater than that for an X-ray 'photograph'.

3 Why are X-ray techniques termed 'non-invasive'?

8.3 The use of ultrasound

At the end of Section 8.3 you should be able to:
- describe the properties of ultrasound
- describe the piezo-electric effect
- explain how ultrasound transducers emit and receive high-frequency sound
- describe the principles of ultrasound scanning
- describe the difference between A-scan and B-scan
- calculate the acoustic impedance using the equation $Z = \rho c$

- calculate the fraction of reflected intensity using the equation

$$\frac{I_r}{I_0} = \frac{(Z_2 - Z_1)^2}{(Z_2 + Z_1)^2}$$

- describe the importance of impedance matching
- explain why a gel is required for effective ultrasound imaging techniques
- explain what is meant by the Doppler effect
- explain qualitatively how the Doppler effect can be used to determine the speed of blood.

The generation of ultrasound

Ultrasound waves are pressure waves similar to sound but with frequencies above those that can be heard by the human ear. Generally, ultrasound is said to have frequencies greater than 20 kHz.

Ultrasound waves may be generated using a **piezo-electric transducer**. A transducer is the name given to any device that converts energy from one form to another. In this case, electrical energy is converted into ultrasound energy by means of a piezo-electric crystal such as quartz.

In a quartz crystal, opposite faces of the crystal become oppositely charged when subject to pressure. Conversely, when subject to an electric field, the crystal changes shape. This is the **piezo-electric effect**.

The structure of quartz is made up of a large number of tetrahedral silicate units, as shown in Figure 8.16. These units build up to form a crystal of quartz that can be represented, in two dimensions, as shown in Figure 8.17.

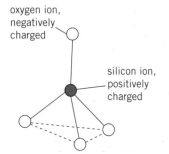

oxygen ion, negatively charged

silicon ion, positively charged

Figure 8.16 Tetrahedral silicate unit

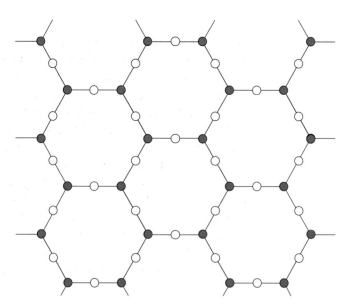

Figure 8.17 Two-dimensional representation of a quartz crystal

When the crystal is unstressed, the centres of charge of the positive and the negative ions in any one unit coincide, as shown in Figure 8.18a.

Electrodes may be formed on opposite sides of the crystal by depositing silver on its surfaces. When a potential difference is applied between the electrodes, an electric field is set up in the crystal. This field causes forces to act on the ions. The oxygen ions are negatively charged and the silicon ions have a positive charge. The ions are not held rigidly in position and, as a result, they will be displaced slightly when the electric field is applied across the crystal. The positive ions will be attracted towards the negative electrode and the negative ions will be attracted to the positive electrode. Dependent on the direction of the electric field, the crystal will become slightly thinner (Figure 8.18b) or slightly thicker (Figure 8.18c).

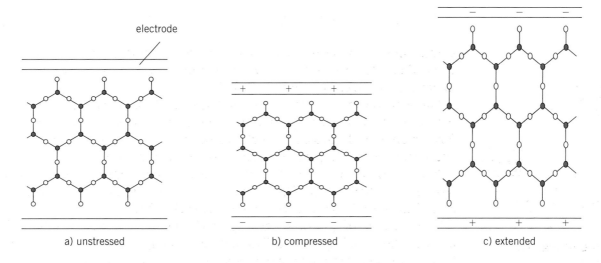

a) unstressed   b) compressed   c) extended

Figure 8.18 The effect of an electric field on a quartz crystal

An alternating voltage applied across the electrodes causes the crystal to vibrate with a frequency equal to that of the applied voltage. These oscillations are likely to have a small amplitude. However, if the frequency of the applied voltage is equal to the natural frequency of vibration of the crystal, resonance will occur and the amplitude of vibration will be a maximum. The dimensions of the crystal can be such that the oscillations are in the ultrasound range of frequencies (greater than about 20 kHz). These oscillations will give rise to ultrasound waves in any medium surrounding the crystal.

If a stress is applied to an uncharged quartz crystal, the forces involved will alter the positions of the positive and the negative ions, creating a potential difference across the crystal. Therefore, if an ultrasound wave is incident on the crystal, the pressure variations in the wave will give rise to voltage variations across the crystal. An ultrasound transducer may therefore also be used as a detector (or receiver).

A simplified diagram of a piezo-electric transducer/receiver is shown in Figure 8.19.

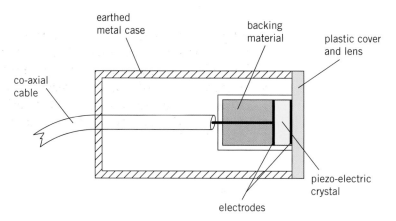

Figure 8.19 Piezo-electric transducer/receiver

A transducer such as this is able to produce and detect ultrasound in the megahertz frequency range, which is typical of the frequency range used in medical diagnosis.

The reflection and absorption of ultrasound

Ultrasound is typical of many types of wave in that, when it is incident on a boundary between two media, some of the wave power is reflected and some is transmitted. This is illustrated in Figure 8.20.

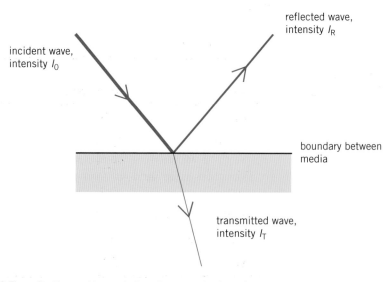

Figure 8.20 The reflection and transmission of a wave at a boundary

For a wave of incident intensity I_0, reflected intensity I_R and transmitted intensity I_T, by conservation of energy,

$$I_0 = I_R + I_T$$

Although, for a beam of constant intensity, the sum of the reflected and transmitted intensities is constant, their relative magnitudes depend not only on the angle of incidence of the beam on the boundary but also on the media themselves. The relative magnitudes of I_R and I_T are quantified by reference to the **specific acoustic impedance** Z of each of the media. This is defined as the product of the density ρ of the medium and the speed c of the wave in the medium. That is,

$$Z = \rho c$$

For a wave incident normally on a boundary between two media having specific acoustic impedances of Z_1 and Z_2, the ratio of the reflected intensity I_R to the incident intensity I_0 is given by

$$\frac{I_R}{I_0} = \frac{(Z_2 - Z_1)^2}{(Z_2 + Z_1)^2}$$

The ratio I_R/I_0 is known as the **intensity reflection coefficient** for the boundary and is given the symbol α. As the above equation shows, α depends on the difference between the specific acoustic impedances of the two media. Some typical values of specific acoustic impedance are given in Table 8.2, together with the approximate speed of ultrasound in the medium.

Table 8.2 Values of speed of ultrasound and specific acoustic impedance for some media

medium	speed/m s^{-1}	specific acoustic impedance/ kg m^{-2}s^{-1}
air	330	430
blood	1600	1.6×10^6
bone	4100	$5.6 \times 10^6 - 7.8 \times 10^6$
fat	1500	1.4×10^6
muscle	1600	1.7×10^6
soft tissue	1600	1.6×10^6
water	1500	1.5×10^6

Example

Using data from Table 8.2, calculate the intensity reflection coefficient for a parallel beam of ultrasound incident normally on the boundary between:

(a) air and soft tissue,

(b) muscle and bone that has a specific acoustic impedance of $6.5 \times 10^6 \text{kg m}^{-2}\text{s}^{-1}$.

(a) $\alpha = (Z_2 - Z_1)^2 / (Z_2 + Z_1)^2$
$= (1.6 \times 10^6 - 430)^2 / (1.6 \times 10^6 + 430)^2$
$= \textbf{0.999}$

(b) $\alpha = (6.5 \times 10^6 - 1.7 \times 10^6)^2 / (6.5 \times 10^6 + 1.7 \times 10^6)^2$
$= \textbf{0.34}$

Now it's your turn

Using data from Table 8.2,

(a) suggest why, although the speed of ultrasound in blood and muscle is approximately the same, the specific acoustic impedance is different.

(b) calculate the intensity reflection coefficient for a parallel beam of ultrasound incident normally on the boundary between fat and muscle.

It can be seen that the intensity reflection coefficient for a boundary between air and soft tissue is approximately equal to unity. This means that when ultrasound is incident on the body, very little is transmitted into the body. In order that ultrasound may be transmitted into the body and also that the ultrasound may return to the transducer, it is important that there is no air between the transducer and the skin (soft tissue). This is achieved by means of a water-based jelly. This jelly has a specific acoustic impedance of approximately $1.5 \times 10^6 \text{kg m}^{-2}\text{s}^{-1}$.

Obtaining diagnostic information using ultrasound

The ultrasound transducer is placed on the skin, with the water-based jelly excluding any air between the transducer and the skin (Figure 8.21). Short pulses of ultrasound are transmitted into the body where they are partly reflected and partly transmitted at the boundaries between media in the body such as fat–muscle and muscle–bone. The reflected pulses return to the transducer where they are detected and converted into voltage pulses. These voltage pulses can be amplified and processed by electronic circuits such that the output of the circuits may be displayed on a screen as in, for example, a cathode-ray oscilloscope.

Pulses of ultrasound are necessary so that the reflected ultrasound pulses can be detected in the time intervals between the transmitted pulses. The time

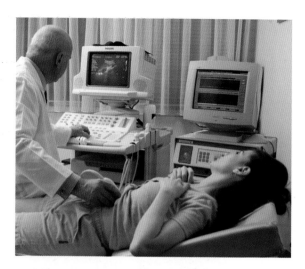

Figure 8.21 Ultrasound diagnosis

between the transmission of a pulse and its receipt back at the transducer gives information as to the distance of the boundary from the transducer. The intensity of the reflected pulse gives information as to the nature of the boundary. Two techniques are in common use for the display of an ultrasound scan.

In an **A-scan**, a short pulse of ultrasound is transmitted into the body through the coupling medium (the water-based jelly). At each boundary between media, some of the energy of the pulse is reflected and some is transmitted. The transducer detects the reflected pulses as it now acts as the receiver. The signal is amplified and displayed on a cathode-ray oscilloscope (c.r.o.). Reflected pulses (echoes) received at the transducer from deeper in the body tend to have lower intensity than those reflected from boundaries near the skin. This is caused not only by absorption of wave energy in the various media but also, on the return of the reflected pulse to the transducer, some of the energy of the pulse will again be reflected at intervening boundaries. To allow for this, echoes received later at the transducer are amplified more than those received earlier. A vertical line is observed on the screen of the c.r.o. corresponding to the detection of each reflected pulse. The time-base of the c.r.o. is calibrated so that, knowing the speed of the ultrasound wave in each medium, the distance between boundaries can be determined. An example of an A-scan is illustrated in Figure 8.22.

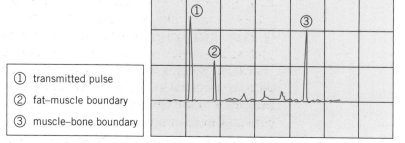

① transmitted pulse

② fat–muscle boundary

③ muscle–bone boundary

Figure 8.22 An A-scan

A **B-scan** consists of a series of A-scans, all taken from different angles so that, on the screen of the c.r.o., a two-dimensional image is formed. Such an image is shown in Figure 8.23.

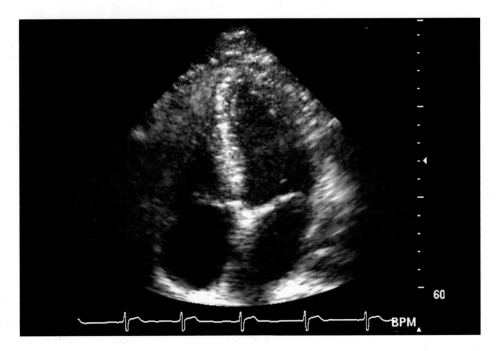

Figure 8.23 An image of a healthy heart produced from a B-scan

The ultrasound probe for a B-scan does not consist of a single crystal. Rather, it has an array of small crystals, each one at a slightly different angle to its neighbours. The separate signals received from each of the crystals in the probe are processed. Each reflected pulse is shown on the screen of the c.r.o. as a bright spot in the direction of orientation of the particular crystal that gave rise to the signal. The pattern of spots builds up to form a two-dimensional image representing the positions of the boundaries within the body. The image may be either viewed immediately or photographed or stored in a computer memory.

The main advantage of ultrasound scanning compared to X-ray diagnosis is that the health risk to both the patient and to the operator is very much less. Also, ultrasound equipment is much more portable and is relatively simple to use.

Higher frequency ultrasound enables greater resolution to be obtained. That is, more detail can be seen. Furthermore, as modern techniques allow for the detection of very low intensity reflected pulses, boundaries between tissues where there is little change in acoustic impedance can be detected.

The Doppler effect

Whenever there is relative motion between a wave source and an observer, then the frequency observed (the apparent frequency) is different to the frequency of the emitted sound. The effect is known as the **Doppler effect**. You may have noticed this effect when a car or a train, sounding its horn, approaches and passes by you. The frequency or pitch of the sound decreases as the source passes you.

Ultrasound is a wave motion and, if there is any motion of the object that reflects the ultrasound, then the frequency detected at the transducer will be different from that emitted. This effect can be used to determine the speed of blood in a blood vessel.

The ultrasound source is stationary and blood in a blood vessel is moving (at an angle) towards the transducer, as shown in Figure 8.24.

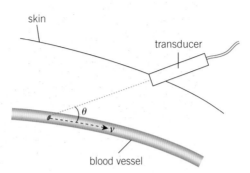

Figure 8.24 Ultrasound and the Doppler effect

The apparent frequency of the ultrasound increases because the reflector (the blood) is moving towards the source (the transducer). There is a second increase in apparent frequency because the blood acts as a source moving towards the detector (the transducer). The total observed change in frequency Δf of the ultrasound of frequency f and speed c is given by

$$\Delta f = \frac{2 f \nu \cos \theta}{c}$$

where ν is the speed of the blood, moving at angle θ to the direction of the incident wave.

By monitoring the change of frequency of the ultrasound of known frequency, it is possible to obtain a value for the average speed of the blood in the vessel.

This technique may be used in other situations where there is movement, for example, monitoring fetal heartbeat.

Section 8.3 Summary

- Ultrasound may be generated and detected by piezo-electric crystals.
- Ultrasound images are formed as a result of the detection and processing of ultrasound pulses that have been reflected from tissue boundaries.
- The acoustic impedance Z of a medium is: ρc
- The intensity reflection coefficient is: $(Z_2 - Z_1)^2 / (Z_2 + Z_1)^2$
- Two-dimensional scans may be obtained using a generator/detector consisting of many separate crystals all at different angles of orientation.

Section 8.3 Questions

1 Explain why, when obtaining an ultrasound scan,
   (a) the ultrasound is pulsed and is not continuous,
   (b) the reflected signal received from deeper in the body is amplified more than that received from near the skin.

2 The specific impedance of fat, muscle and bone are $1.4 \times 10^6\,kg\,m^{-2}\,s^{-1}$, $1.6 \times 10^6\,kg\,m^{-2}\,s^{-1}$ and $6.5 \times 10^6\,kg\,m^{-2}\,s^{-1}$ respectively.
   A parallel beam of ultrasound of intensity I is incident on a boundary. Discuss quantitatively, in terms of I, the reflection and the transmission of the beam of ultrasound as it passes through the boundaries between
   (a) the fat and muscle
   (b) the muscle and bone

3 (a) State what is meant by the Doppler effect.
   (b) Suggest why, when measuring blood flow rate in a blood vessel using the Doppler effect in ultrasound, the speed of the blood is an average value.

8.4 Radioactive tracers

At the end of Section 8.4 you should be able to:
- describe the use of medical tracers like technetium-99m to diagnose the function of organs
- describe the main components of a gamma camera
- describe the principles of positron emission tomography (PET).

Radioisotopes may be used as tracers for diagnostic purposes. Basically, a radionuclide is incorporated within a pharmaceutical product that is then administered to the patient. Since the radionuclide will be giving off radiation, the path of the radioactive material (the tracer) through the body can be followed.

The radioactive tracer should not interfere with any of the functions of the body, otherwise diagnostic information will not reflect what is actually happening within the patient. Also, the tracer must emit radiation of a type and energy that enables it to be detected outside the body. Furthermore, the half-life of the radioisotope should be chosen so that its activity is high enough for the diagnosis to take place but that it does not remain within the body for an excessive length of time.

One long-established application of radioactive tracers is the use of iodine-132 to monitor the thyroid gland. The thyroid gland produces thyroxine, for which it requires iodine. Iodine-132 in a solution of sodium iodide will be absorbed into the thyroid gland to a greater extent than surrounding tissues. This enables the size of the gland to be assessed. Furthermore, the rate of uptake of the iodine, and its rate of excretion, enables over-activity or under-activity of the gland to be assessed.

Technetium-99m is now frequently used as an alternative to iodine-132. It is taken up by the thyroid gland in the same way as iodine. However, it is released more rapidly and therefore the radiation dose is reduced.

Some radioisotopes and their applications are listed in Table 8.3.

Table 8.3 Some radioisotopes used in medicine

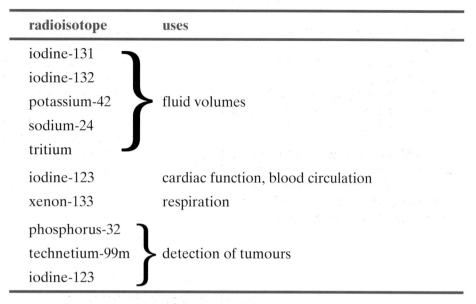

radioisotope	uses
iodine-131	
iodine-132	
potassium-42	fluid volumes
sodium-24	
tritium	
iodine-123	cardiac function, blood circulation
xenon-133	respiration
phosphorus-32	
technetium-99m	detection of tumours
iodine-123	

The radiation emitted by radioisotopes has to be detected. One device used to monitor γ-ray emissions is the gamma camera.

The gamma camera

The gamma camera operates on the same basic principle as a **scintillation counter**. When materials such as anthracene, naphthalene and sodium iodide absorb a γ-ray photon, they may emit a photon of visible light. This process is known as scintillation and the material is referred to as being a fluorescent material.

However, a single photon emitted from the fluorescent material is insufficient to be detected by the human eye. The photon is passed into a photomultiplier tube where it is made to fall on a photocathode, emitting an electron by the photoelectric effect. This electron can then be accelerated towards a dynode at a positive potential with respect to the photocathode. The dynode is a material which, when hit by the electron, emits several electrons. These electrons are then accelerated towards another dynode at a higher positive potential, resulting in further electrons being emitted. After

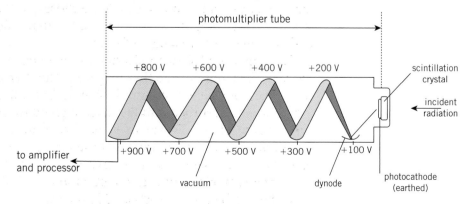

Figure 8.25 A scintillation counter

205

this process has been repeated along a series of about ten dynodes, the single original electron has resulted in an 'avalanche' of electrons that can be amplified and processed. The rate of current pulses will be proportional to the intensity of the γ-radiation incident on the original fluorescent material. A scintillation counter is illustrated in Figure 8.25.

The gamma camera can be considered to be many scintillation counters all grouped together. This is illustrated in Figure 8.26.

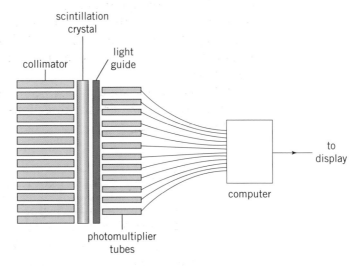

Figure 8.26 A gamma camera

γ-ray photons from the patient travel along a collimator. This collimator consists of a lead grid so that only photons travelling parallel to the holes in the grid will reach the scintillation crystal. Those scintillations produced in the direction of the incident γ-ray photon pass through a light guide and on to an array of photomultipliers. The output from the photomultipliers is fed into a computer where they are processed to give a display.

An example of an image produced by a gamma camera is shown in Figure 8.27.

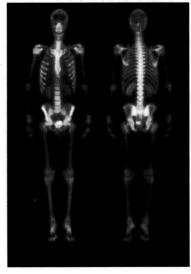

Figure 8.27 A gamma camera scan

Positron emission tomography (PET)

Positron emission tomography (PET) enables three-dimensional images of functional processes in the body to be obtained.

Some radioactive nuclei decay by the emission of positrons. A positron is a particle that has the same mass as an electron. Its charge is the same magnitude as that on an electron but it is positive. When the positron is emitted by the decaying nucleus it loses energy and, within a short distance, it is annihilated by combination with an electron. The annihilation gives rise to two γ-ray photons, each of energy 0.51 MeV, that are produced simultaneously. To conserve momentum, these photons move off in opposite directions.

PET relies on the simultaneous detection of two γ-ray photons that are travelling in opposite directions. Equipment for PET is shown in Figure 8.28.

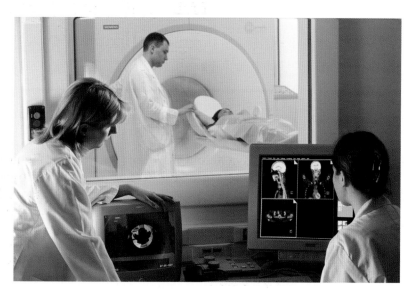

Figure 8.28 Equipment for obtaining a PET scan

The γ-ray detectors are arranged in a circle around the patient. Pairs of γ-ray photons detected in opposite directions within a few nanoseconds of each other are recognised. Others are ignored. The source of the positron must lie along the line joining the two detectors. The outputs from the detectors are coupled to a computer where the signals are stored and processed. A typical PET scan is shown in Figure 8.29.

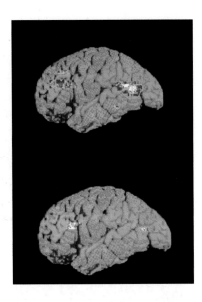

Figure 8.29 A PET scan

The sources usually used have short half-lives. Some of these sources are listed in Table 8.4.

Table 8.4 Some radioisotopes used in PET scans

radioisotope	half-life/minutes
carbon-11	20
nitrogen-13	10
oxygen-15	2
fluorine-18	110

The radioisotope to be used is introduced into the body on a metabolically active molecule such as glucose. The scan will enable those regions of the body that are metabolically active to be monitored. The technique is frequently used to diagnose tumours, in neurology and in cardiology.

Example

Suggest and explain four factors that must be considered when choosing a suitable radioisotope for use in diagnosis.

1 The tracer must not be poisonous or affect bodily functions as otherwise the diagnosis could be erroneous.

2 The emitted radiation must have appropriate energy so that it can be detected from outside the body.

3 The radioisotope (tracer) must be capable of being used as a label for an appropriate pharmaceutical product so that it can be administered to the patient and the part of the body to be examined will take up the tracer.

▶▶

4 The half-life of the radioisotope must be long enough so that the diagnostic procedure may be undertaken but short enough to prevent over-exposure of the patient to unnecessary radiation.

Now it's your turn

Diagnosis using positron emission tomography (PET) relies on the detection of γ-ray photons.

State and explain the necessary conditions for the photons to be selected for processing to produce the PET scan.

Section 8.4 Summary

- Radioisotopes have many uses in medicine including diagnosis of tumours, the assessment of fluid volumes and the functioning of organs.
- Radioisotopes provide a non-invasive means of diagnosis.
- In general, diagnosis involves the ingestion by the patient of the radioisotope and the subsequent detection, outside the body, of the emitted radiation.
- The radioisotope is used to label molecules chosen so that, when taken up by the body, the function of the organs is not affected.
- Various techniques are used for the detection of the emitted radiation including GM tubes, scintillation counters in various arrays (the gamma camera). Data collected from the detectors is stored on computer and analysed to produce images. These images can be coloured to assist subsequent diagnosis.

Section 8.4 Questions

1 (a) What is meant by a radioisotope?

(b) Explain why diagnosis involving radioisotopes imposes a risk on the patient.

2 Discuss why, although radioisotopes pose a risk to the health of the patient, their use is justified.

3 Suggest why, in many diagnostic techniques such as using a gamma camera and in PET scans, the development of these techniques could not take place until powerful computers were made available.

8.5 The use of (nuclear) magnetic resonance imaging (MRI)

At the end of Section 8.5 you should be able to:

■ outline the principles of magnetic resonance, with reference to precession of nuclei, Larmor frequency, resonance and relaxation times
■ describe the main components of an MRI (magnetic resonance imaging) scanner
■ outline the use of MRI to obtain diagnostic information about internal organs
■ describe the advantages and disadvantages of MRI.

Nuclear magnetic resonance

Many atomic nuclei have a property that is known as 'spin'. The 'spin' causes the nuclei to behave as if they were small magnets. Such nuclei have an odd number of protons and/or neutrons. Examples of these nuclei are hydrogen, carbon and phosphorus.

When a magnetic field is applied to these nuclei, they tend to line up along the field. However, this alignment is not perfect and the nuclei rotate about the direction of the magnetic field, due to their spin. The motion can be modelled as the motion of a top spinning about the direction of a gravitational field. This rotation is known as **precession**.

The spinning about the direction of the magnetic field has a frequency known as the frequency of precession or the **Larmor frequency**. The frequency depends on the nature of the nucleus and the strength of the magnetic field.

If a pulse of electromagnetic radiation of the same frequency as the Larmor frequency is incident on precessing nuclei, the nuclei will resonate, absorbing energy. This frequency is in the radio-frequency (RF) band with a wavelength shorter than about 10 cm. A short time after the incident pulse has ended, the nuclei will return to their equilibrium state, emitting RF radiation. The short time between the end of the RF pulse and the re-emitting of the radiation is known as the **relaxation time**. The whole process is referred to as **nuclear magnetic resonance**.

In practice, there are two relaxation processes and it is the time between these two that forms the basis of **magnetic resonance imaging (MRI)**. Since hydrogen is abundant in body tissues and fluids, hydrogen is the atom used in MRI.

Magnetic resonance imaging (MRI)

A schematic diagram of a magnetic resonance scanner is shown in Figure 8.30.

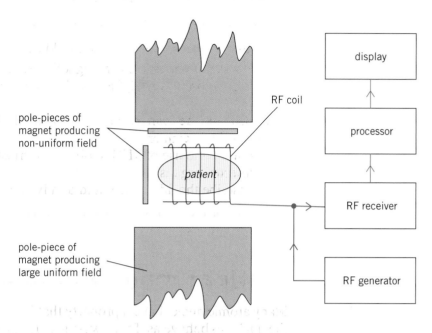

Figure 8.30 Schematic diagram of a magnetic resonance scanner

The patient who is to be investigated is positioned in the scanner (see Figure 8.30), between the poles of a large magnet that produces a very large uniform magnetic field (in excess of 1 tesla). The magnetic field causes all the hydrogen nuclei within the person to precess with the same Larmor frequency. In order that the hydrogen nuclei in only one small part of the body may be detected, a non-uniform magnetic field is also applied across the patient. This non-uniform field is accurately calibrated and results in a different magnitude of magnetic field strength at each point in the body of the patient. Since the Larmor frequency is dependent on the strength of the magnetic field, the Larmor frequency will also be different in each part of the patient. The particular value of magnetic field strength, together with the radio-frequency that is emitted, enables the hydrogen nuclei in the part under investigation to be located.

Radio-frequency (RF) pulses are produced in coils near the patient. These pulses pass into the patient. The emitted pulses produced as a result of de-excitation of the hydrogen nuclei are picked up by the coils. These signals are processed and displayed so as to construct an image of the number density of hydrogen atoms in the patient. As the non-uniform magnetic field is changed, atoms in different parts of the patient's body are detected and displayed. An image of a cross-section through the patient may be produced. One such image is shown in Figure 8.31.

A whole series of images of sections through the patient can be produced and stored in a computer memory. As with CAT scanning, this enables a three-dimensional image to be generated. Sections through the patient may be viewed from many different angles.

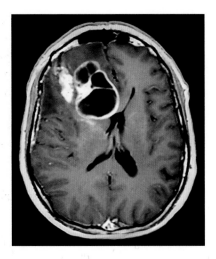

Figure 8.31 MRI scan showing a brain tumour

Magnetic resonance imaging (MRI) is expensive, both as regards the initial cost of the equipment and the running costs. Furthermore, the equipment is massive and cannot be moved easily. Since the patient must lie within a large magnet, some develop a feeling of claustrophobia. It is necessary for the patient to remain motionless for a significant length of time. This is possible for adults but young children must be sedated.

Despite these disadvantages, MRI is now a widely-used technique. The images are three-dimensional and can be rotated to obtain a view from any angle. The detail obtained is superior to most other techniques.

Example

The use of magnetic resonance imaging is now a common diagnostic tool in medicine. However, there may be disadvantages for some patients. Suggest why:
(a) some people may find the scanner to be claustrophobic,
(b) there are additional problems when scanning young children rather than adults.

(a) In order to produce a large magnetic field through the patient, the patient must be placed inside a large cylindrical magnet. This can mean that the patient feels trapped.
(b) The total scan takes some time because separate scans of many slices through the patient must be taken. The patient must be very still whilst this is done and this is difficult for children.

Now it's your turn

During an MRI scan, the patient is placed in a large uniform magnetic field on to which is superimposed a highly calibrated non-uniform magnetic field. Explain the purpose of each of these two magnetic fields.

Section 8.5 Summary

- Magnetic resonance imaging (MRI) monitors the concentration of hydrogen nuclei in the body.
- Hydrogen nuclei have 'spin' which causes then to precess in an applied magnetic field.
- The frequency of precession (the Larmor frequency) depends on the magnetic field applied to the nuclei.
- A radio-frequency pulse at the Larmor frequency causes the nuclei to resonate
- On de-excitation, the nuclei give off radio-frequency waves that can be detected and analysed.

Section 8.5 Questions

1 Explain briefly what is meant by nuclear magnetic resonance.

2 (a) By reference to nuclear magnetic resonance imaging, explain what is meant by

  (i) the Larmor frequency,
  (ii) relaxation time.

(b) Suggest, with an explanation, a typical value for the Larmor frequency.

Exam-style Questions

1 CAT scanning and MR scanning provide images of sections through the patient.
 (a) Describe, by reference to what is detected in each case, the differences between these two types of scan.
 (b) Suggest why both types of scan rely on the use of powerful computers.

2 (a) By reference to an X-ray image, explain what is meant by *sharpness* and *contrast*.
 (b) Describe how sharp X-ray images of soft organs, such as the stomach, may be obtained.
 (c) The linear attenuation (absorption) coefficients of X-rays in muscle and in bone are $0.89\,\text{cm}^{-1}$ and $3.0\,\text{cm}^{-1}$ respectively. In a particular limb, a bone of thickness 1.5 cm is surrounded by muscle of thickness 3.0 cm. A parallel beam of X-rays is incident on the limb.
 Calculate the fraction of the incident intensity that is transmitted through the limb.

3 (a) Describe the principles behind the use of magnetic resonance to obtain diagnostic information about internal body structures.
 (b) Suggest three reasons why MR scanning would not be used to diagnose a simple bone fracture.

9. Modelling the universe

The study of astrophysics and cosmology deals with the largest objects in the universe – stars and galaxies. This is in stark contrast to our work on nuclear physics and fundamental particles in Chapter 7. Nevertheless, some of the ideas and processes we learnt about there are of direct relevance to a study of the universe. In this chapter, we show how theories of the origin of the universe have been developed as experimental evidence has been refined, and discuss possible origins of the universe and try to predict how it may continue to develop in the future.

9.1 Stars, galaxies and radiation

At the end of Section 9.1 you should be able to:

- describe the principal contents of the universe, including stars and galaxies
- describe the solar system in terms of the Sun, planets, planetary satellites and comets
- understand that planets in the solar system are detected by the sunlight they reflect
- understand that stars and galaxies are detected by the electromagnetic radiation they emit
- define the following units of distance used in astronomy: astronomical unit (AU), light year (ly) and parsec (pc), and use these units in calculations.

The term **astronomy** is used to describe the study of celestial bodies in all their aspects. Until about 1900 work was mainly of an observational nature, making use of physical tools such as optical telescopes, spectrographs and photographs to obtain quantitative information on the position of heavenly bodies and of the spectra and intensity of the radiation emitted by them. Early astronomy also included the collection of a vast database of calculation and computation directed towards predicting the positions of planets and their satellites. The application of physical ideas to explain the processes involved in the creation and subsequent life of stars and other heavenly bodies is covered in the area of astronomy called **astrophysics**. **Cosmology**, the study of the universe as a whole and the construction of a theoretical framework to understand its origin and its future, is based on Einstein's general theory of relativity, and in particular its theory of gravitational forces.

Stars, galaxies and radiation: an introduction

According to the ancient astronomers and philosophers, our Earth was the centre of the universe. It was the centre of a system of nesting transparent spheres, which rotated round the Earth at different rates. Heavenly bodies, such as the Sun, our Moon and the planets, were fixed on these spheres. The so-called fixed stars were on the surface of the outermost, stationary sphere, somewhere beyond the outermost planet. The universe was thus neatly self-contained and of fixed size.

We now have a very different picture of the universe. The Earth-centred model of the universe put forward by the ancients has long been abandoned. The Earth is one of a number of planets moving round a star we call the Sun in approximately circular orbits. The Earth, the planet so familiar to us, is the third from the Sun. Some of the planets, including Earth, have one or more satellites; we call our satellite the Moon. We know that the Sun is only one of a very large number of stars (making some very rough assumptions, we shall show that it may be about 10^{11}) in a Galaxy called the Milky Way. The Milky Way is only one of a large number, a universe, of galaxies. The largest telescopes have been able to collect information from a number of galaxies outside our own. It has been estimated that there are at least 10^{10} galaxies.

Imagining that we could view our own Galaxy from outside, we would see, in side view, a disc-shaped conglomeration of stars with a bulging central nucleus, and in plan view spiral arms coming out from the nucleus, as sketched in Figure 9.1. There is nothing special about our Sun; it is just one of the very many stars in the Galaxy, located more than halfway from the nucleus to the edge. The Sun orbits the centre of the Galaxy (the galactic centre) with a period of about 200 million years.

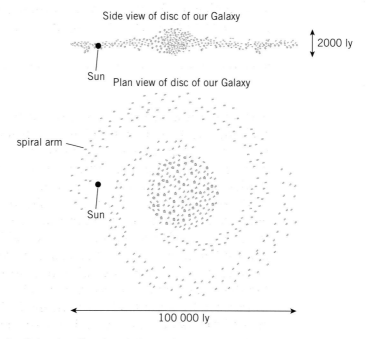

Figure 9.1 Our Galaxy is a disc-shaped cluster of stars

How do we obtain information about the universe? For ground-based astronomy, there are only two transparent windows in the absorption spectrum of the Earth's atmosphere. These are a band from the near ultraviolet, through the visible to the near infrared region of the electromagnetic spectrum (covering a wavelength range from about 300 nm to 1 μm) and another band covering the microwave, UHF and short-wavelength radio wave regions (wavelengths from about 0.1 m to 100 m). Observations at other parts of the electromagnetic spectrum require telescope instrumentation outside the atmosphere, to avoid absorption of the radiation. Some early work was done with rocket-mounted telescopes and spectrometers, but this suffered from the disadvantage that the period available for observation was very short, being restricted to the interval between the rocket passing through the atmosphere and then returning to Earth. A great quantity of spectacular work of high quality has been done in the last twenty years with telescope instrumentation on spaceships and satellites.

Traditionally, by the use of optical telescopes, ground-based astronomers have for centuries made observations of stars and galaxies by the visible light they emit. In addition to information about the positions of stars, the colour of light emitted can be used to give information about the star's temperature and age.

Visible light is only a very small part of the radiation making up the electromagnetic spectrum. A great deal of information has been obtained since about 1930 through picking up radiation from another part of the electromagnetic spectrum, radio waves. This opened up a new field in astronomy, radio astronomy, with new observational techniques. These include the use of single large dish receivers and linear arrays of aerials. One wavelength of interest is 21 cm, corresponding to a frequency of 1.4×10^9 Hz, which is really in the radar rather than radio wave region. The 21 cm radiation comes from electrons in the hydrogen atoms in clouds of the gas in space; it does not come from a transition between energy levels that we have already met in the hydrogen atom but from another type of electron energy transition. Both proton and electron in a hydrogen atom spin about their axes; they can do this so that their spins are either parallel (both clockwise or both anticlockwise) or antiparallel (one clockwise and the other anticlockwise). These spin configurations have different energies. When the electron switches from the parallel spin mode to the antiparallel, a photon of energy of 5.9×10^{-6} eV is emitted. This is the energy of a photon of 21 cm radiation.

Another part of the spectrum of great cosmological importance is the microwave region. In 1964 Robert Wilson and Arno Penzias were annoyed by a constant background noise in their radio telescope. The intensity of the noise really was constant; it did not depend on time of day or season of the year, or on direction. It was found to have a peak wavelength of 1.1 mm. It could be deduced that this radiation came from outside our galaxy, in fact from all parts of the universe. The value of the peak wavelength suggested that the radiation came from a source at a temperature of 2.7 K (sometimes rounded to 3 K). This radiation is known as the **cosmic microwave background** (or CMB).

Example

Our Galaxy can be modelled as a disc with a central bulge (see Figure 9.1). Most of the mass of the Galaxy is in this bulge. Our Sun (and solar system) orbits the centre of the Galaxy with a speed of about $2.5 \times 10^5\,\text{m s}^{-1}$ and at a distance from the centre of about $2.8 \times 10^{20}\,\text{m}$. Estimate the mass of the Galaxy and the number of stars in it.

We'll start by calculating the mass M of the Galaxy's central bulge. Since this is where most of the mass is concentrated, it can be approximated as a point mass and we can apply exactly the same ideas as we did in considering the orbits of the planets of the solar system (see Section 3.2). By realising that the centripetal force was provided by the gravitational attraction, we found that

$$GMm/r^2 = mv^2/r.$$

If we apply the same equation to the Galaxy, we have

$$M = rv^2/G$$

where r is the radius of the solar system's orbit about the centre of the Galaxy and v is its speed.

When we substitute the numerical values, we have

$$M = (2.8 \times 10^{20})(2.5 \times 10^5)^2/6.7 \times 10^{-11} \approx \mathbf{3 \times 10^{41}\,kg}.$$

If all the Galaxy's stars were to have a mass approximately equal to that of our Sun ($2.0 \times 10^{30}\,\text{kg}$), this means that there would be very roughly $3 \times 10^{41}/2.0 \times 10^{30}$, or in an order-of-magnitude calculation, $\mathbf{10^{11}}$ stars in it. Note that in calculations about the universe we shall very often have to make some sweeping assumptions about the magnitudes of quantities.

Astronomical distances and units

A most important date in the history of astronomy was 1609, when Galileo constructed his first telescope and began to take observations of planets and stars. Almost at once it was realised that the universe was very much larger than had been previously thought, and that the Earth was not at its centre. It was found that distances, even within the solar system, were very large.

To measure lengths in units that are appropriate in a terrestrial laboratory or on the ground is not so manageable when it comes to measuring the distances of stars, or even other planets, from the Earth. Thus the distance between the Earth and the Moon is $3.8 \times 10^8\,\text{m}$, which can just about be grasped. But the Earth–Sun distance is $1.5 \times 10^{11}\,\text{m}$, which is too large to handle easily. The nearest star to us, apart from the Sun, is Proxima Centauri, which is about $4 \times 10^{16}\,\text{m}$ away. Astronomers use three other units in measuring distances in the universe. For measurements within the solar system, the **astronomical unit** (AU) is very convenient.

The **astronomical unit** is defined as the radius of the circular path round the Sun followed by a body in 365.25 days.

From your knowledge that the length of a year is 365.25 days, you will recognise that the astronomical unit is the mean distance between Earth and Sun, and is thus equal to 1.5×10^{11} m. This means that the distances between the various planets and the Sun vary between 0.39 AU for Mercury and 39.5 AU for Pluto, with Earth, by definition, at 1.00 AU. The Earth–Moon distance is 2.6×10^{-3} AU.

Distances between stars are much greater; Proxima Centauri is about 2.7×10^5 AU away from the Sun. For such large distances, it is convenient to express the measurement in terms of the time it takes light to travel this distance.

The **light year** (ly) is the distance traversed by electromagnetic radiation in one year.

Using the relation speed = distance/time, it is easy to show that one light year is equal to 9.46×10^{15} m or 6.33×10^4 AU. This makes Proxima Centauri about 4.3 ly away from the Sun. Note that a light year is not a unit of time!

Another unit for large astronomical distances is called the **parsec**, derived from the words *par*allax and *sec*ond.

The **parsec** is the distance from which the radius of the Earth's orbit subtends an angle of one second of arc.

Since the radius of the Earth's orbit is one astronomical unit, an alternative definition is

The **parsec** is the distance from which one astronomical unit subtends an angle of one second of arc.

The definition of the parsec is illustrated in Figure 9.2.

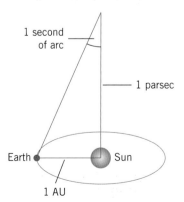

Figure 9.2 Definition of the parsec (not to scale)

Using the definition of the radian and the conversion factors between radians and degrees and degrees and seconds, it is easy to show that 1 parsec is equal to 2.06×10^5 AU. Thus, it is also equal to 3.26 light years or 3.09×10^{16} m.

Example

The parallax angle of a certain star is $0.000\,150\,°$. How far away is the star? Give the answer in parsecs and in light years.

This is a straightforward question testing knowledge and understanding of units, and conversion between them.

$0.000\,150°$ is equal to $0.000\,150 \times 60 \times 60 = 0.540$ seconds of arc. The distance of the star, in parsecs, is then $1/0.540 = $ **1.85 pc**.

Since 1 pc = 3.26 ly, the distance can also be expressed as **6.03 ly**.

Now it's your turn

A certain star is 20 pc away. How long does it take its light to reach us?

The solar system

We have already mentioned that the Earth is one of a number of planets moving round the Sun in approximately circular orbits. These planets, with their own satellites (such as our Moon), are part of what is called the solar system. The force that keeps the solar system together, and which provides the centripetal force for the orbital motion of the planets, is the gravitational force. The nine planets are listed in Table 3.1, together with their orbital radii r and periods T. We saw in Section 3.2 how Newton's inverse square law of gravitation follows from Kepler's third law of planetary motion, that is, T^2 is proportional to r^3.

In addition to the nine well-known planets and their satellites, the solar system contains a number of small or dwarf planets. One of these, Ceres, orbits the Sun between Mars and Jupiter; the others are beyond Pluto. Ceres is in a belt of many very small orbiting rocky bodies called the asteroid belt. (see Figure 9.3.)

Another, more spectacular, class of member of the solar system are the comets. These have orbits which are markedly elliptical rather than approximately circular, with the Sun at one focus of the ellipse. This ellipticity means that the comets return regularly, often passing close enough to the Earth to be visible with the naked eye. Comets are bright objects with a distinctive tail, giving the impression that they are perhaps self-propelled by the expulsion of burning gases. However, comets are made of ice in which

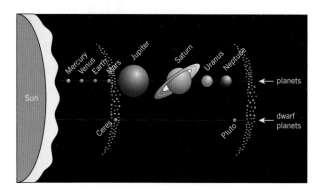

Figure 9.3 The planets of the solar system

Figure 9.4 Part of the Bayeux Tapestry showing Halley's Comet. The Latin text says 'they stare amazed at the star'.

interstellar dust and fragments of rock are embedded; they are generally only a few kilometres in diameter, and are perhaps best thought of as being similar to dirty snowballs. The bright light we see from a comet is reflected light from the Sun, and the tail, which points away from the Sun, is merely vapour.

Note that the components of the solar system – planets and their satellites, and comets – are all observed by light reflected from the Sun. This is in contrast to stars and galaxies, where the information we receive comes from the electromagnetic radiation they emit.

Section 9.1 Summary

- Information about the universe is obtained using optical and radio telescopes to detect electromagnetic radiation.
- The universe contains more than 10^{10} galaxies.
- Our Sun and the solar system are part of the Milky Way Galaxy, which contains about 10^{11} stars.
- The solar system contains nine major planets and a number of dwarf planets which orbit the Sun in approximately circular orbits, and comets, which have markedly elliptical orbits.
- The major planets obey Kepler's third law: T^2 is proportional to r^3, where T and r are the orbital period and radius respectively.
- Astronomical distances are measured in astronomical units ($1\,\text{AU} = 1.5 \times 10^{11}\,\text{m}$), light years ($1\,\text{ly} = 9.46 \times 10^{15}\,\text{m}$) and parsecs ($1\,\text{pc} = 3.09 \times 10^{16}\,\text{m}$).

Section 9.1 Questions

1 In cosmology we have been obliged to invent new units for distance: the light year and the parsec.
   (a) Explain why this was necessary. Why does it not seem so necessary to invent a new unit for time?
   (b) Explain to a person with only limited knowledge of physics that the light year is a unit of distance, not of time.

2 The word *galaxy* comes from two Greek words meaning way of *milk*. In view of the fact that the universe contains a vast number of galaxies and clusters of galaxies, was this choice of name appropriate?

9.2 The evolution of the Sun

At the end of Section 9.2 you should be able to:

- describe the formation of a star, such as our Sun, from interstellar dust and gas
- describe the Sun's probable evolution into a red giant and white dwarf
- describe how a star much more massive than our Sun will evolve into a super red giant and then either a neutron star or a black hole.

The main sequence of stars

There is a great range of brightness in the stars that are visible to us. Some are mere pinpoints of light, whereas the star we know most about, our Sun, is much brighter than any other object in the heavens. It is easy to think of two factors responsible for these differences in brightness: the quantity of energy emitted by a star, and the distance of the star from the observer on Earth. Another factor is the surface temperature of the star. The early astronomers were able to detect differences in the colour of stars. The colour of a star is related to its surface temperature. Reddish stars are dimmer and cooler than bluish stars. Furthermore, surface temperature is related to the mass of the star. In general, the coolest (reddest) stars have low mass. (By cool, we mean a surface temperature in the region of 3500 K; the hottest stars may have temperatures up to 50 000 K, and these are distinctly blue.) The brightest and hottest (most blue) stars have high mass. Our Sun, with a surface temperature of about 6500 K, is yellowish–white, at the cooler end of the range.

This describes the general correlation between colour, surface temperature and stellar mass. Stars that fit this pattern (the great majority of stars) are called **main sequence** stars.

Red giants and white dwarfs

There are also stars that fall outside the main sequence classification. For example, stars that are both very bright and very massive but have low (red) surface temperatures are called **red giants**. Some other stars are dim but have a very high surface temperature, hence appearing white; these are called **white dwarfs**.

Stellar evolution – the Sun

Why are there stars of different temperatures (and colours)? Will they retain their colour characteristics throughout their lifetimes? The answer is no; different colours represent different stages in the evolution of a star. In this section we will follow the formation and evolution of a star such as our Sun.

Although one thinks of empty space as being a perfect vacuum, there is a very small amount of gas in it. We are talking of only one or two gas molecules for every hundred cubic metres of space. Most of this gas is molecular hydrogen, with some helium atoms. The temperature of space is very low indeed, so that the molecules move slowly. Although the overall density of gas is very small, there are regions where the concentration is higher, and a cloud of gas builds up. If the concentration becomes sufficiently high the clouds of hydrogen molecules contract due to the gravitational force between them. The molecules will accelerate towards the centre of mass of the cloud. As the kinetic energy of the molecules increases, the temperature of the cloud increases. At a temperature of a few thousand kelvin, the bonds between hydrogen atoms break and the atoms lose their electrons, creating a hydrogen plasma. The increasing kinetic energy of the particles will eventually reach a point where the electrostatic repulsion between positively-charged hydrogen nuclei in the plasma is overcome so that nuclear fusion can take place. A cloud of gas large enough to create a star by fusion is called a **proto-star**.

In Section 7.11 we have already met one of the fusion reactions that takes place in a star like our Sun. This involved the fusion of two deuterium (hydrogen-2 nuclei) to give a helium-3 nucleus and a neutron, with the release of energy. Other fusion reactions between hydrogen isotopes take place as well. All of these reactions require the fusing nuclei to have very considerable kinetic energies, of the order of 10^{-16} J. This corresponds to a temperature in the gas cloud of the order of 10^7 K. Such high temperatures are found only in the core of the proto-star. The enormous quantity of energy released by fusion produces an outwards pressure sufficient to stop the gravitational contraction, and the proto-star reaches a state of equilibrium. This stable state will place the young star somewhere on the main sequence. Exactly where it fits on the main sequence will depend on the mass of the gas cloud when the process started.

As the hydrogen fusion reaction in the core proceeds, the denser helium nuclei formed will accumulate in the central core of the star, where they were produced. Hydrogen nuclei in the shell around the core continue to react. When helium gathers in the core, the core contracts because helium is heavier than hydrogen. The acceleration of hydrogen nuclei in the shell outside the core is boosted, their temperature increases, and the fusion reaction proceeds even more fiercely. The outer envelope of the star expands and cools somewhat. This reduction in surface temperature makes the star move towards the red end of the main sequence. Eventually its increase in size will make the star much brighter and it will move off the main sequence to enter the red giant stage. Our Sun is at present on the main sequence; it has been there for about 4.5×10^9 years. It will probably stay on the main sequence, but becoming redder all the time, for another 5×10^9 years, before moving to the red giant stage. And why do we call these stars *giant*? The expansion of the star as it moves from the main sequence to the red giant stage is very considerable. It has been estimated that the Sun will expand from its current size to a diameter roughly that of the Earth's orbit, that is an expansion by a factor of about 200.

As the star expands, its core shrinks and reaches even higher temperatures. At a core temperature of about 10^8 K, helium nuclei, which have twice the charge of hydrogen nuclei and hence four times the electrostatic repulsion, have enough energy to undergo fusion. The main product of the helium fusion reaction is carbon-12 (proton number 6), but further reactions can produce nuclei of even greater mass. At stellar core temperatures of 5×10^9 K nuclei as heavy as iron-56 and nickel-56 can be synthesised by fusion. This process of nucleosynthesis, the formation of heavy nuclei from light ones by fusion, is limited by the fact that for a nucleon number of about 56 the binding energy per nucleon curve has reached its peak (see Figure 7.12). Further fusions would not release energy, but require it.

For stars of mass about that of the Sun, no further energy can be obtained from fusion reactions in the red giant stage, and the star will shrink under the action of its internal gravitational attraction. As it gets smaller, its emitted radiation becomes whiter. The star has now evolved into a white dwarf. Why call these stars *dwarf*? Approximate calculations show that when the Sun reaches the white dwarf stage, it will be about the size of the Earth. Its density will increase so much that its diameter will decrease to about one-hundredth of its present size. Eventually the star will be able to shrink no more, as its atoms will be so close together that their electron clouds start to overlap. It will continue to cool down by the loss of the internal energy accumulated by the fusion reactions of the red giant stage, so that it will become dimmer and dimmer until it is nothing but a lump of burnt ash called a **black dwarf**.

The life story of the Sun can be summarised as follows. It starts as a proto-star. The ions in the hydrogen plasma of which it is made start to fuse, and the young Sun reaches equilibrium somewhere in the main sequence. Increasing the rate of the hydrogen fusion reactions causes an increase in size but a cooling of the surface, so that the Sun will move towards the red giant stage. Helium fusion reactions then take over, eventually causing the formation of products of higher and higher nucleon number, but up to a limit of about 60. No further fusion energy can be obtained and the Sun will become a white dwarf. When all its internal energy is exhausted, it will be burnt out, finishing up as a black dwarf.

Example

A student reads that the energy output of the Sun is due to the following sequence of fusion reactions:

$$^1_1H + {}^1_1H \rightarrow {}^2_1H + e^+ + \nu \qquad (0.42\,\text{MeV})$$

$$^1_1H + {}^2_1H \rightarrow {}^3_2He + \gamma \qquad (5.49\,\text{MeV})$$

$$^3_2He + {}^3_2He \rightarrow {}^4_2He + {}^1_1H + {}^1_1H \quad (12.86\,\text{MeV})$$

where ν represents a neutrino and the energy released in each reaction is stated in brackets.

To obtain the energy released in the overall sequence

$$4{}^1_1H \rightarrow {}^4_2He + 2e^+ + 2\nu + 2\gamma$$

the student adds up all the figures in brackets to obtain the figure of 18.77 MeV. He is surprised to read on and find that, in fact, the energy released in the sequence is 26.72 MeV. Suggest explanations for the discrepancy.

First of all, the student has made an elementary mistake in misreading the cycle. It takes *two* of each of the first two reactions to produce the two ${}^3_2He$ needed for the third reaction, so the total energy output appears to be $(2 \times 0.42) + (2 \times 5.49) + 12.86 = 24.68\,\text{MeV}$. But this is still not quite right. A second discrepancy has occurred because the positron (e^+) produced in each of the first reactions annihilates very quickly with an electron (e^-) producing an extra $2m_ec^2 = 1.02\,\text{MeV}$ of energy per reaction, giving a total of $24.68 + (2 \times 1.02) = 26.72\,\text{MeV}$. This is the overall energy release.

Now it's your turn

In the red giant stage, two important fusion reactions are

$$^4_2He + {}^4_2He \rightarrow {}^8_4Be + \gamma$$

$$^4_2He + {}^8_4Be \rightarrow {}^{12}_6C + \gamma$$

The second fusion quickly follows the first, giving an overall effect of

$$3{}^4_2He \rightarrow {}^{12}_6C$$

Calculate the energy released in the overall reaction.

(Nuclear masses: ${}^4_2He$, 4.001 502 u; ${}^{12}_6C$, 11.9967 u)

Stellar evolution – massive stars

Stars of mass much greater than that of the Sun follow a different route. The great mass of the star allows its core to collapse even further under gravity than the normal limit for a red giant, that is the limit set by the maximum of the binding energy per nucleon curve. The kinetic energy of the nuclei, and thus the temperature of the core of this super red giant, is so high that the fusion of elements of greater nucleon number than iron takes place, even though these reactions require energy. Also, collisions between particles of such high energy can cause the breaking apart of iron and nickel nuclei into helium nuclei, and hence to protons and neutrons. There is so much energy available that some electrons and protons can be forced together to produce neutrons, in a sort of inverse β-decay reaction. The core of the star contracts more and more until it consists almost entirely of neutrons. We now have a **neutron star** of extremely high density. The contraction of the core continues until the neutrons are so close together that they can be compressed no more. The sudden stoppage of the contraction produces a shock wave that may cause the whole star to explode violently, producing what is called a **supernova**. A great deal of energy is emitted during the supernova explosion, blowing away the entire outer envelope of the star as debris into space. The energy of the explosion is sufficient to form virtually all the elements of the Periodic Table; some of this debris is thought to be the origin of the heavy elements in the solar system. Other remnants of the supernova may be observed as a **nebula**. Supernovas certainly exist; a famous example of a very intense star, which then suddenly disappeared, was recorded by Chinese astronomers in 1054. The Crab nebula (Figure 9.5) contains the remains of the explosion.

Figure 9.5 The Crab nebula

There is another possibility for a massive neutron star. Here the gravitational force on the neutrons in the contracting core becomes so strong that it prevents light emitted from the star from escaping. Effectively, the light is pulled back by the force of gravity. If no radiation is emitted, the star cannot be seen; it appears black. If another body were to approach it too closely, it would be swallowed up by the **black hole**. Black holes are predicted by theory, and there is some observational evidence suggesting that they do exist.

Section 9.2 Summary

- A star begins life as a collapsing cloud of hydrogen. It is then in the proto-star stage.
- The collapse of the hydrogen cloud causes an increase in temperature of the gas as potential energy is transformed into kinetic energy. When the temperature reaches about $10^7 \, \text{K}$, nuclear fusion occurs and helium is produced, followed by heavier elements.
- The energy produced during fusion balances the gravitational energy so that the young star stabilises as a main sequence star.
- Eventually, as the hydrogen in the stellar core is consumed, helium collects there; the core contracts and heats up further while the

envelope expands and cools. The star moves off the main sequence and becomes a red giant.

■ For a star like the Sun, the next stage is further cooling of the envelope and the star becomes a white dwarf. Finally, it will cool down completely until it ceases to radiate energy and is a black dwarf.

■ A star heavier than the Sun contracts even further because of the greater gravitational forces. Eventually it becomes an enormous nucleus made up of neutrons (a neutron star). The energy produced from the final contraction of the core is released as a supernova explosion.

■ Even heavier stars may contract even further to produce such a strong gravitational field that no radiation can escape. This is a black hole.

Section 9.2 Questions

1 A star generates energy by nuclear fusion reactions at its core. How does it lose this energy so that it comes to a state of equilibrium? What happens to the equilibrium when the rate of generation of energy gets less?

2 How 'giant' is a red giant? What effect does the size of a red giant have on its properties?

3 In what sense is a black hole black? In what sense is it a hole?

9.3 The expanding universe

At the end of Section 9.3 you should be able to:

■ state Olbers' paradox
■ interpret Olbers' paradox to explain why it suggests that the model of an infinite, static universe is incorrect
■ select and use the equation $\Delta\lambda/\lambda = v/c$
■ describe and interpret Hubble's redshift observations
■ state and interpret Hubble's law
■ convert the Hubble constant H_0 from its conventional units $(\text{km s}^{-1}\text{Mpc}^{-1})$ to SI units (s^{-1})
■ state the cosmological principle
■ describe and explain the significance, in relation to the cosmological principle, of the 3 K cosmic microwave background radiation.

Olbers' paradox

We have just discussed how stars go through an evolution cycle from proto-star, main sequence star, red giant to white dwarf, or to neutron star and black hole. The fact that individual stars evolve suggests that the universe as a whole evolves, rather than remaining static.

Newton's picture of the universe was that it was static, that is, it did not change over time. But there are difficulties in assuming that the universe is static and finite. The main problem is that a static, finite universe is unstable.

Gravitational forces should pull all its components together to produce an enormous lump of matter at the centre.

A static, infinite universe also presents problems. One of these was put forward by the German astronomer Olbers in 1823. Everyone agrees that the night sky is dark. Yet, if the universe were infinite, along any line of sight there would be a star emitting light towards the Earth. So the night sky ought to appear uniformly bright. It doesn't, so the implication is that the universe is finite. But this seems to be impossible, because of the stability argument.

Hubble and the Doppler shift

The answer to the paradox came with a suggestion from Edwin Hubble in 1923. This was that distant galaxies are moving away from us, and that their speed is proportional to their distance away. That is, the universe is expanding. This hypothesis immediately puts paid to the idea of a static universe.

The experimental evidence for Hubble's suggestion came from previous observations that the wavelengths of identifiable spectral lines in the spectra of light from distant galaxies did not correspond with the wavelengths as measured in the laboratory on Earth. The colour of the light seemed to be shifted towards the red end of the spectrum. Observation of this **red shift** led to one of the most important ideas in modern theories of the universe.

To explain a shift in the colour of light, either towards the red end or the blue end of the spectrum, we need to recall an effect that has long been familiar in relation to the frequency and wavelength of a source of sound if the source is moving towards or away from the observer. You will have noticed that the pitch of the siren of a police car or ambulance drops as it passes you. This is because the vehicle, as it moves towards you, catches up with wavefronts previously emitted from the siren. You, as a stationary observer listening to the note from the approaching siren, will detect more wavefronts in unit time; that is, the frequency is higher (or the wavelength is longer). As the vehicle passes you and moves away, the wavefronts are further apart than normal. Fewer wavefronts than normal are detected in unit time, and the frequency is lower. This phenomenon is called the **Doppler effect**.

The situation is slightly different in the case of light because of relativity; there is then no distinction between motion of the source and motion of the observer. The detailed derivation of the equation for the Doppler effect has to take this into account. But making a slight approximation – that the velocity of the moving source is rather less than the speed of light – the equation for the small change in wavelength $\Delta\lambda$ is

$$\Delta\lambda/\lambda \approx \pm v/c \quad \text{(for } v \gg c\text{)}$$

or, in words, the fractional change in wavelength is proportional to the speed of the source of light towards or away from the observer. The sign of the change in wavelength is such that for motion away from the observer, the wavelength is increased. That is, light emitted from a source moving away from an observer has its wavelength shifted towards the red – it has undergone a *red shift*. The red shift effect is illustrated in Figure 9.6. (We talk about a red shift, although this term can be ambiguous. An *increase* in wavelength in a microwave source moving away from us is still described as a

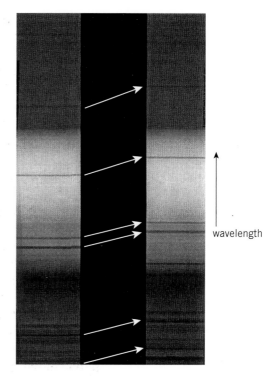

wavelength

Figure 9.6 The red shift.
The strip on the left shows absorption lines in the spectrum of the Sun, and the strip on the right the same absorption lines in the spectrum of light from a receding galaxy. Wavelength increases upwards, so the light from the galaxy has been red-shifted.

red shift, although if the shift were really towards the red this would correspond to a *decrease* in wavelength.)

A red shift in the spectra of some stars had been observed as early as the mid-nineteenth century. Hubble made detailed measurements on the spectra of light from many galaxies, and found that the red shift effect was general. The suggestion is that the universe is expanding; the galaxies in it are moving away from each other. This is a most important observation. The universe cannot be static.

Another important outcome of Hubble's work was the observation that the amount of the red shift of light from a particular galaxy is approximately proportional to the distance of the galaxy from us. Applying the Doppler equation, where we found that the shift in wavelength is proportional to the speed of the source, we obtain **Hubble's law** as

$$v \approx H_0 d$$

where v is the velocity of a galaxy moving away from us, d is its distance away, and H_0 is a constant called the Hubble constant.

The 'approximately equal to' sign in the equation is there because the law is not precise. In fact, some nearby galaxies are moving towards us, and hence their spectra are blue-shifted. This is believed to be due to random motion among the galaxies. For very distant galaxies, the speed of recession is much larger than that of the random motion of the nearer ones, and there can be no doubt that the universe is expanding. The value of the Hubble constant H_0 is taken to be roughly $50 \, \text{km s}^{-1} \text{Mpc}^{-1}$. This value can only be approximate because the distances d for very remote galaxies are not certain.

Note that the unit $\text{km s}^{-1} \text{Mpc}^{-1}$, which is the one in conventional use because the distances of galaxies are so large, can readily be converted to SI units.

We have $1 \, \text{km} = 10^3 \, \text{m}$ (the metre is the SI base unit of distance)

and $1 \, \text{Mpc} = 3.1 \times 10^{22} \, \text{m}$ (see page 219)

so

$$1 \, \text{km s}^{-1} \text{Mpc}^{-1} = 10^3 \, \text{m s}^{-1}/3.1 \times 10^{22} \, \text{m} = \mathbf{3.2 \times 10^{-20} \, s^{-1}}.$$

We have been talking as if we on Earth are acting as the observational centre of the universe. We have said that galaxies are moving away from us, implying that the Earth is the point at the centre of what might have been an enormous explosion a very long time ago. But this is not the case. The expansion of the universe would appear to be the same from any other observing point in the universe. This is consistent with an important and very basic assumption in cosmology. The **cosmological principle** states that, on a large scale, the universe is both isotropic and homogeneous. In more simple language, the universe looks the same in all directions and would look the same if we were to observe it from another galaxy. Considered locally, the cosmological principle does not apply. For example, we can identify different constellations of stars in the night sky by looking in different directions, so on our own small scale the universe is not isotropic. But if we look on a large enough scale the average population density of galaxies ought to be the same in whichever direction we choose. Another factor in favour of the cosmological principle is the uniformity in both time and direction of the cosmic microwave background (CMB) radiation, which we met on page 216.

However, there are some observations that throw doubt on the applicability of the cosmological principle. There is no doubt that galaxies tend to cluster, so can the universe really be homogeneous? A solution that has been proposed by some cosmologists is to suppose that the universe might be composed of up to 90% of non-luminous dark matter which is uniformly distributed throughout the universe, thus preserving its homogeneity.

The CMB radiation has an importance far beyond that of helping to back up the cosmological principle. This importance relates to the support it gives to a particular theory of the evolution of the universe, the **Big Bang** theory, to which we will return in Section 9.4.

Example

A certain galaxy is receding from us at a speed of approximately $3 \times 10^6 \, \text{m s}^{-1}$.

(a) Estimate the distance of the galaxy from us.
(b) Estimate the observed wavelength of the 656 nm line in the Balmer series of hydrogen in the spectrum of light from this galaxy.

(a) Here we use Hubble's law to find the distance. We will take the Hubble constant as $50 \, \text{km s}^{-1} \text{Mpc}^{-1}$, as this is merely an 'estimate' question. Applying the law, we have

$$v = H_0 d,$$

so $d = v/H_0 = 3 \times 10^3\,\mathrm{km\,s^{-1}}/50\,\mathrm{km\,s^{-1}\,Mpc^{-1}} = \mathbf{60\,Mpc}$

Note the conversion from $3 \times 10^6\,\mathrm{m\,s^{-1}}$ to $3 \times 10^3\,\mathrm{km\,s^{-1}}$.

(b) This part is an application of the Doppler equation

$$\Delta\lambda/\lambda = v/c$$

or $\Delta\lambda = \lambda v/c = (656 \times 10^{-9}\,\mathrm{m})(3 \times 10^6\,\mathrm{m\,s^{-1}})/3 \times 10^8\,\mathrm{m\,s^{-1}}$
$$= 656 \times 10^{-7}\,\mathrm{m} \approx 7\,\mathrm{nm}$$

Since the galaxy is receding, the change in wavelength is a redshift and the observed wavelength is $656 + 7 = \mathbf{663\,nm}$.

Now it's your turn

Estimate the speed of a galaxy that is about 10^7 light years away from us. (This galaxy will be close to the edge of the universe.)

Section 9.3 Summary

■ Olbers' paradox states that if the universe is infinite and static, the night sky should be uniformly bright; it is not, so this model is incorrect.

■ The Doppler effect gives the fractional change in wavelength for the wavelength of light emitted from a source moving with speed v as $\Delta\lambda/\lambda = v/c$. If the source is receding from the observer this corresponds to a wavelength shift towards the red end of the spectrum.

■ Hubble noted that, in general, light from galaxies was red-shifted, and interpreted this as a continuous expansion of the universe.

■ Hubble's law states that the velocity of recession v of a galaxy is approximately proportional to its distance d from the observer, or $v \approx H_0 d$, where H_0 is the Hubble constant.

■ The value of the Hubble constant is about $50\,\mathrm{km\,s^{-1}\,Mpc^{-1}} = 3.2 \times 10^{-14}\,\mathrm{s^{-1}}$.

■ The cosmological principle states that the universe is both isotropic and homogeneous.

■ The fact that the 2.7 K (3 K) cosmic microwave background (CMB) radiation is independent of direction supports the cosmological principle.

Section 9.3 Questions

1 Some people feel that all the important advances in cosmology (the Big Bang theory, the discovery of the cosmic microwave background radiation and so on) have been made in the last 60 years or so. List what you consider to be the important advances that were made before 1950.

2 Why do we use such a complicated unit for the Hubble constant ($\mathrm{km\,s^{-1}\,Mpc^{-1}}$) when the SI unit ($\mathrm{s^{-1}}$) is much less cluttered?

3 All galaxies appear to be moving away from us. A friend says that this means that the Earth must be at the very centre of the universe. Explain to him why he is wrong.

9.4 The evolution of the universe

At the end of Section 9.4 you should be able to:
- describe and explain the significance, in relation to the evolution of the universe, of the cosmic microwave background (CMB) radiation
- explain that the hot Big Bang model of the universe implies a finite age for the universe
- select and use the expression age of universe $\approx 1/H_0$
- describe qualitatively the evolution of the universe from 10^{-43} s after the Big Bang to the present
- explain that the universe may be 'open', 'flat' or 'closed', depending on its density
- explain that the ultimate fate of the universe depends on its density
- define the term *critical density*
- select and use the expression for the critical density of the universe $\rho_0 = 3H_0^2/8\pi G$
- explain that it is currently believed that the density of the universe is close to, and possibly exactly equal to, the critical density needed for a 'flat' cosmology.

Introduction to the Big Bang

We hinted in the last section that the expansion of the universe started with an enormous explosion. We now develop this idea, the **Big Bang** theory, to provide a detailed picture of the evolution of the universe.

What is the evidence for such a huge explosion as the starting point of the universe? We shall see that the idea of an expanding universe allows us to calculate the age of the universe from the time when the Hubble expansion began. This age fits with data about the age of stars, and with the age of the oldest rocks, obtained from radioactivity experiments. Another crucial piece of evidence came from the discovery of the cosmic microwave background (CMB) radiation.

An important feature of this radiation was its uniformity in time and direction; we have seen that this was vital in supporting the cosmological principle. But it also helps us to understand what the Big Bang was like. To produce the whole universe, and to cause it to expand, must have required a very great release of concentrated energy. The temperature must have been so high that there could not have been any identifiable atoms. The universe must have consisted only of elementary particles and radiation. The existence of the CMB radiation suggests that matter, in the form of elementary particles, and radiation were once in equilibrium at a very high temperature indeed. The photons of the radiation were trapped by the particles, particularly by electrons; as soon as the photons were emitted they were scattered or absorbed. As the universe expanded, its energy was spread out over a larger volume of space, causing a reduction in temperature. It might have taken about 3×10^5 years for the temperature to fall to 3000 K, and below this temperature some nuclei and electrons could have remained in the atomic, rather than the ionic, state. As the free electron density got less there was less scattering and absorption of radiation, so the radiation could spread through the universe. As the universe expanded, the radiation underwent a continuous increase in wavelength towards the red; it was red-shifted. (Note that this is not the radiation from stars and galaxies; it is the background radiation.) Longer wavelengths correspond to lower temperatures; we are at present at a state where the wavelength of the background is characteristic of a temperature of 2.7 K.

The total energy associated with the CMB radiation is very much greater than the total radiation energy from all the stars in the universe. However, it is small compared with the energy associated with the mass of the universe (from $E = mc^2$). We can say that the universe is matter-dominated. But in the early stages of the universe's life, before much matter had been created, it was certainly radiation-dominated by the energy of the CMB radiation.

We have seen that all the experimental facts – CMB radiation, Hubble expansion and radioactivity measurements – fit in with the idea of the Big Bang occurring at a certain defined time. This allows us to state with confidence that the universe has a finite age.

The age of the universe

The idea of Hubble expansion is based on the assumption that at one time all the galaxies were close together. As a result of the Big Bang, they all started to move apart with speeds that increased with their distance. If we date the Big Bang to the time when expansion started, we can very easily obtain an estimate of the age of the universe.

The value of the Hubble constant in SI units is $H_0 = 50 \times 3.2 \times 10^{-20}\,\text{s}^{-1} = 1.6 \times 10^{-18}\,\text{s}^{-1}$. (It is much more convenient to carry out this calculation in SI units rather than in the conventional units of $\text{km}\,\text{s}^{-1}\,\text{Mpc}^{-1}$.) To reach the present separation of the galaxies from zero separation would require a time of

$$t_0 = d/v = 1/H_0 = 1/1.6 \times 10^{-18}\,\text{s}^{-1} = 6.3 \times 10^{17}\,\text{s} \approx 2 \times 10^{10} \text{ years}.$$

This value, the reciprocal of the Hubble constant, is sometimes called the **Hubble age**, or more formally the **characteristic expansion time**. This is probably an overestimate of the age of the universe. Galaxies have not been moving with unchanged speeds, as they are slowed down by their mutual gravitational attraction. Working from the Hubble equation, a better estimate might be between 1.0×10^{10} years and 1.8×10^{10} years.

We have mentioned other estimates of the age of the universe. One comes from a determination of the age of the Earth (and hence of the solar system) from the radioactivity of our oldest rocks. The estimate here is about 4.5×10^9 years. This is likely to be less than the age of the universe because the solidification of rocks on Earth must have taken place a considerable time after the formation of the universe. Another estimate is from the theory of stellar evolution, giving a value for the age of stars of between 1×10^{10} years and 1.5×10^{10} years. So the estimate, based on the Hubble age, of between 1.0×10^{10} years and 1.8×10^{10} years, seems to be acceptable.

The evolution of the universe from 10^{-43}s to the present

We now review in a qualitative manner the life of the universe from its creation, or at least very shortly after the Big Bang took place. This account will summarise what is called the **Standard Cosmological Model**.

We start at a time of about 10^{-43}s after the Big Bang. Why 10^{-43}s? We do not have a theory at present to account for what happened earlier than this time. The pressures, densities and temperatures at earlier times require an idea of quantum effects on gravity; such a theory is simply not available to us. Even at 10^{-43}s the temperature was unimaginably high, about 10^{32}K, and any particles would have been moving with an average kinetic energy of 10^{19}GeV. The four forces in nature (strong nuclear, weak nuclear, electromagnetic and gravitational) could not be differentiated from each other.

At 10^{-43}s a change occurred in which the gravitational force appeared as a separate force. However, there was no distinction between quarks and leptons, and baryon and lepton numbers were not conserved. As time passed to about 10^{-35}s the universe had expanded, and the temperature had fallen to 10^{27}K. At this stage the strong nuclear force could be distinguished, as could quarks and leptons. The leptons included electrons, neutrinos and others, and all the corresponding antiparticles. The quarks were initially free, but soon they began to form nucleons and the other hadrons, and of course the corresponding antiparticles. Particles and antiparticles were continually annihilating, but the energy available was sufficient to create them.

As time passed further to about 10^{-6}s after the Big Bang, the temperature fell to about 10^{13}K and the average kinetic energy of particles was now insufficient to create particle–antiparticle pairs from nucleons. Annihilation did continue, however. This situation led to a universe with far fewer nucleons than earlier, and with (for a reason that is not quite clear) more particles than antiparticles.

Some minutes after the Big Bang, the temperature was about 10^9K. Nuclear fusion began to take place. The kinetic energy of the colliding protons and neutrons was sufficient for nuclei such as deuterium and helium to be formed by fusion, but not so large that the newly created elements would be shattered by subsequent collisions. The universe was cooling so rapidly that the synthesis of heavier nuclei stopped, so that after about an hour of existence the matter in the universe was mainly hydrogen nuclei (protons), helium nuclei and free electrons. Radiation (photons) dominated the situation, as we described in the section on the CMB radiation.

After a long time (on the scale of the tiny fractions of a second we have been thinking about so far) had passed – perhaps 3×10^5 years after the Big Bang – the temperature was only about 3000K. This was low enough for the nuclei present to capture electrons and retain them in orbit, forming neutral atoms. The background radiation continued to decrease in energy, until it corresponded to a temperature of only a few kelvin. But the energy of the universe due to the mass of its particles did not decrease, and the universe became matter-dominated.

After the creation of atoms, stars and galaxies began to form by gravitational attraction. This would have been about 10^6 years after the Big Bang. The process continues today, 1.0×10^{10} years to 1.8×10^{10} years later.

This description is a vastly simplified summary of the Standard Cosmological Model. The Model has been tweaked several times since it was first put forward. One of the more important tweaks (sometimes called the **inflationary theory**) relates to the introduction of a very rapid exponential expansion of the universe at about 10^{-35}s after the Big Bang. After this brief, rapid increase in extent the universe followed its normal pattern of expansion. This modification is helpful in explaining some features of the curvature of space–time, to which we will refer in the next part.

Will the universe expand forever?

We know from the Standard Cosmological Model that the universe is continuing to evolve and expand. Will this expansion continue without limit?

To answer this question we have to know something about the curvature of space–time. The theory of relativity requires not merely the three-dimensional system of spatial coordinates (x-, y- and z-coordinates) we use in classical (Euclidean) solid geometry, but a four-dimensional system that includes time as well as space.

We are certainly familiar with plane geometry; all the geometrical theorems we use are based on simple axioms such as the one about the angles of a triangle adding up to 180°. But this is only applicable if the triangle we are considering is drawn with straight lines on a flat, plane sheet of paper. If you draw a triangle on the surface of a sphere, its sides will not be straight lines and the angles will not add up to 180°. (If the triangle on the spherical surface is very small, its sides will approximate to being straight because the small piece of the sphere on which it is drawn is very nearly plane. The sum of the angles is then approximately 180°.) The discrepancy between plane and spherical geometry has arisen because we have moved from a concept of flat space to the idea of a curved space.

This is easy enough to understand so far. But it gets much more difficult to comprehend in the four-dimensional space–time system required by Einstein in the theory of relativity, particularly when the theory is applied to such a large system as the universe which contains very massive bodies. A simplified picture for space in the universe is that of a thin, stretched rubber sheet. This will normally be a plane surface. But suppose we place a heavy mass, representing a galaxy, on the sheet. It will sag, causing space to curve. A body moving in the neighbourhood of the heavy mass, or a passing ray of light, will tend to move along a curved line near the depression in the rubber. (Incidentally, this provides a rather good analogy to the behaviour of a black hole, which we saw on page 224 is so dense that matter, and even light, cannot escape from it.)

What effect does the curvature of space have on the universe on a large scale, not just in a region near the heavy-mass-on-a-rubber sheet, or black hole, analogy? We need to decide whether space has positive or negative curvature, or is flat. (To illustrate, a sphere is said to have positive curvature, but a saddle-shaped surface is said to have negative curvature at the dip of the saddle. A plane surface is said to have, rather obviously, zero curvature.) If space has positive curvature, it is finite, or *closed*. An over-simplistic interpretation of this statement would be to take it as meaning that the universe goes out to a certain distance, and then just stops. It is not as easy as this; in a closed universe, space folds back and returns to the origin. If space has negative or zero curvature it is an *open* universe that just goes on and on; it is infinite. Whether the universe is open or closed depends on the total mass of the universe, and hence on its density.

We can now try to answer the question about whether the universe will continue to expand forever. Figure 9.7 shows the three different possibilities for the future size of the universe, which depend on whether it is open or closed, or whether its curvature is negative, zero or positive.

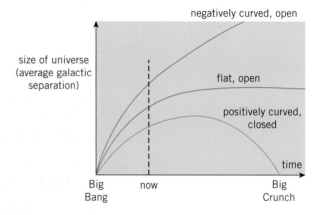

Figure 9.7 Three possibilities for the future of the universe

If space–time in the universe has *negative* curvature or is flat, expansion will not stop, although the rate of expansion will slow down. In the case of flat space–time, the rate of expansion will reach zero. In both of these situations the universe is open and infinite. The third curve shows what happens for space–time with *positive* curvature, which is for a closed and finite universe. Here the effect of gravity will be so strong (because the universe contains so much matter and has such a high density) that expansion would stop and the universe would begin to contract. All the matter in the universe would come back together in what has been called, in analogy to the Big Bang, the **Big Crunch**. So the ultimate fate of the universe depends on the amount of matter in it, or on the density of the universe.

The critical density of the universe

Clearly, the big question is whether our universe is following one of the upper two curves in Figure 9.7, or that corresponding to the curve for a closed universe. As we have seen, it all depends on the density of the universe. The **critical density**, ρ_0, is the density corresponding to the universe with no curvature of space–time, the middle one of the three curves. An expression for the critical density of the universe is

$$\rho_0 = 3H_0{}^2/8\pi G$$

Estimates of the actual density of the universe have produced values one or two orders of magnitude less than the critical density, suggesting that the universe is open. However, there is also evidence for the existence of a significant quantity of non-luminous matter (or **dark matter**) in the universe. One example comes from studies of the rotation of galaxies. They seem to rotate as if they had more mass than we can see. Also, the inflationary expansion theory suggests that the density of the universe should be exactly equal to the critical density.

So, although the answer to our question is by no means definite, it seems possible that the density of the universe is close to the critical density needed for a flat cosmology. The universe is thus open and infinite. The expansion of the universe will never stop, but will continue closer and closer towards a limit.

Examples

1 Find the numerical value (in $kg\,m^{-3}$) of the critical density of the universe. Estimate the number density of nucleons to which this corresponds.

Substituting in the expression $\rho_0 = 3H_0{}^2/8\pi G$, we have

$$\rho_0 = 3 \times (1.6 \times 10^{-18}\,s^{-1})^2/8\pi \times 6.7 \times 10^{-11}\,N\,m^{-2}\,kg^{-2}$$
$$\approx \mathbf{5 \times 10^{-27}\,kg\,m^{-3}}$$

The mass of a nucleon is $1.7 \times 10^{-27}\,kg$, so the critical density corresponds to a number density of nucleons of
$5 \times 10^{-27}/1.7 \times 10^{-27} \approx \mathbf{3\,m^{-3}}$

2 Show that, in spherical geometry, the angles of a triangle may not add up to 180°.

We will take the simple example of a spherical triangle bounded by the edges of an octant of a sphere (Figure 9.8).

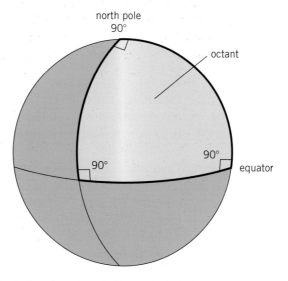

Figure 9.8 An octant of a sphere, bounded by a spherical triangle

Here the angle at the north pole of the sphere is 90°. The other sides are arcs of the circumference of the sphere which run from the pole to the equator. Both of these arcs meet the equator at 90°. In this case, each of the three angles of the triangle is 90°, so the sum of the three angles of this spherical triangle is 270°.

Now it's your turn

Construct a timeline summarising the main features of the evolution of the universe from a time of 10^{-43} s after the Big Bang to the present day.

Section 9.4 Summary

- The existence of the 2.7 K cosmic microwave background (CMB) radiation supports the Standard Cosmological Model of the universe, which is based on the idea of the universe starting with an enormous explosion (the Big Bang).
- The Standard Cosmological Model implies that the universe has a finite age.
- The age of the universe may be estimated by taking the reciprocal of the Hubble constant to give a value which is probably in the range 1.0×10^{10} years to 1.8×10^{10} years.
- During the evolution of the universe, matter became increasingly organised, in the sequence Big Bang – chaos – leptons and free quarks – nucleons (protons and neutrons) and electrons – helium nuclei – hydrogen and helium atoms – star formation – galaxy formation – clusters of galaxies.
- The future of the universe depends on the curvature of space–time. If the universe is open or flat it will continue to expand, but if it is closed the expansion will cease and the universe will eventually contract to a Big Crunch. The deciding factor is whether the average density of the universe is less than, equal to or more than the critical density.
- The critical density is given by the expression $\rho_0 = 3H_0^2/8\pi G$, which has a numerical value of about 5×10^{-27} kg m^{-3}, corresponding to a number density of nucleons of about 3 m^{-3}.
- Estimates lead to the belief that the density of the universe is close or equal to the critical density, implying a flat cosmology and an open universe in which expansion will continue more and more slowly towards a limit.

Section 9.4 Questions

1 Compare an explosion on Earth, for example one such as occurs during the blasting of rock in a quarry, with the Big Bang. How does the speed of the rock fragments as they spread out from the source of the explosion compare with the speed of the proto-stars and stars as they spread out after the Big Bang? Do the rock fragments come to rest? Do the stars and galaxies come to rest? What type of universe does the quarry explosion correspond to, open or closed? Why?

2 It has been suggested that, in a closed universe, the Big Crunch might be followed by a Big Bounce (another Big Bang). What might follow on after this? Could this provide an answer to the question, 'What happened before the Big Bang?'?

3 Recently, there was a misprint in the British Telecom Telephone Directory in which the Department of Physics and *Astronomy* of a certain university was listed as the Department of Physics and Astrology. Staff in the department suffered the inconvenience of taking calls from members of the public who were seeking horoscopes. But, apart from this nuisance factor, did the astronomy and cosmology researchers have any valid cause for complaint?

Exam-style Questions

1 (a) State the cosmological principle.
  (b) Describe the important properties of the cosmic microwave background (CMB) radiation and how the Standard Cosmological Model of the universe explains these properties. Explain their significance as evidence for the past evolution of the universe.
  (c) Explain why our understanding of the very earliest moments of the universe is unreliable.

2 (a) The future of the universe may be *open*, *closed* or *flat*. Explain the meaning of the terms in italics, using a graph to illustrate your answer.
  (b) The mean density of the universe ρ_0 is thought to be approximately $1 \times 10^{-26} \, \text{kg} \, \text{m}^{-3}$. Calculate a value for the Hubble constant H_0.

3 The measure of the size of the universe (the average distance between galaxies) at any moment is believed to have been inversely proportional to the kelvin temperature. Consider the following times after the Big Bang:
  (i) $10^5 \, \text{yr}$, **(ii)** $10^{-6} \, \text{s}$, **(iii)** $10^{-35} \, \text{s}$.
  (a) State the approximate temperatures at each of the times (orders of magnitude will suffice).
  (b) Deduce the measure of the size of the universe at each of the times, compared with the value S of the measure today.

Answers

Chapter 1

Section 1.1 (page 3)

1 1100 kg
2 (a) $7.0 \, \text{m s}^{-1}$
   (b) $4540 \, \text{m s}^{-2}$
   (c) 430 N
3 Yes. The sum of the forces in any direction must be zero.
4 Approximately 150 g. The weight is 1.5 N.

Now it's your turn (page 6)

600 N

Now it's your turn (page 9)

1 $6.8 \times 10^{-24} \, \text{kg m s}^{-1}$
2 $1.2 \, \text{m s}^{-1}$

Now it's your turn (page 10)

$220 \, \text{m s}^{-1}$

Now it's your turn (page 13)

1 (a) $u_A/2$ (b) 0.5

Sections 1.2–1.3 (page 14)

1 $3.3 \times 10^4 \, \text{kg m s}^{-1}$
2 $3.6 \times 10^7 \, \text{N}$
3 $2.7 \times 10^{-4} \, \text{N}$
4 $1.03 \times 10^5 \, \text{m s}^{-1}$
5 Heavy particle's speed is practically unchanged; light particle moves with speed $2u$, in the same direction as the incident heavy particle.
6 the lighter body

Exam-style Questions (page 15)

1 (a) (i) $2.16 \, \text{kg m s}^{-1}$ (ii) $1.07 \, \text{m s}^{-1}$ (iii) 1.16 J
   (b) 0.059 m
2 $3080 \, \text{m s}^{-1}$
4 (a) $3u$ to left   (b) (i) $3u$ to right   (ii) $3t_1$
   (c) $3u/2$
5 (a) $3m$ (b) 0.25

Chapter 2

Now it's your turn (page 19)

1 $9.1 \times 10^{22} \, \text{m s}^{-2}$
2 (a) $7.8 \, \text{km s}^{-1}$ (b) $5.3 \times 10^3 \, \text{s}$

Now it's your turn (page 20)

(a) $0.22 \, \text{m s}^{-1}$ (b) $0.63 \, \text{rad s}^{-1}$ (c) $0.14 \, \text{m s}^{-2}$

Now it's your turn (page 24)

(a) 590 m   (b) 4200 N

Section 2.1 (page 24)

2 (a) $3.3 \, \text{rad s}^{-1}$
   (b) 0.53 revolutions per second

Now it's your turn (page 30)

(a) $x = 0.10 \cos \pi t$ (b) 0.20 s

Now it's your turn (page 33)

(a) 0.25 m   (b) 2.5 s

Sections 2.2–2.4 (page 34)

1 (a) 220 Hz   (b) $1.4 \times 10^3 \, \text{rad s}^{-1}$ (c) $42 \, \text{m s}^{-1}$
   (d) $5.9 \times 10^4 \, \text{m s}^{-2}$ (e) $7.3 \, \text{m s}^{-1}$
2 $2.8 \, \text{m s}^{-1}$
3 $9.826 \, \text{m s}^{-2}$
4 (a) $5.8 \, \text{N m}^{-1}$ (b) 15 mm
   (c) 0.58 s   (d) -8.2 mm

Now it's your turn (page 36)

(a) 0.87 J   (b) 1.97 J   (c) 2.84 J

Section 2.5 (page 37)

1 $3.2 : 1$
2 (a) $0.87A$ (b) $0.71A$

Section 2.6 (page 42)

(a) 2.0 Hz
(b) Suggestion is incorrect; driving frequency will act against natural frequency every half-cycle.

Exam-style Questions (page 43)

1 (b) $2.72 \times 10^{-3} \,\mathrm{m\,s^{-2}}$
2 (a) (i) $25 \,\mathrm{rad\,s^{-1}}$ (ii) 4.0 revolutions per second
   (b) (i) 0.84 N
3 (b) (i) $2A\rho g\Delta x$ (ii) $2g\Delta x/L$
   (iv) $(1/2\pi)\,\sqrt{(2g/L)}$
4 (a) 1.3 s   (b) 16 N
5 (a) $1.28 \times 10^{14} \,\mathrm{Hz}$

Chapter 3

Now it's your turn (page 47)

$25 \,\mathrm{N\,kg^{-1}}$

Now it's your turn (page 51)

1.41 hours

Sections 3.1–3.2 (page 53)

1 $5.5 \times 10^{3} \,\mathrm{kg\,m^{-3}}$
2 $7.78 \times 10^{8} \,\mathrm{km}$
3 -0.50%; -0.25%

Exam-style Questions (page 53)

1 (a) $5.7 \times 10^{-11} \,\mathrm{N}$
2 (b) (ii) $8.86 \times 10^{4} \,\mathrm{km}$

Chapter 4

Now it's your turn (page 58)

$103.6 \,\Omega$

Now it's your turn (page 59)

$1.001 \times 10^{3} \,\mathrm{Pa}$

Section 4.1 (page 60)

1 296 K, 15 K

Now it's your turn (page 65)

1 $3.3 \times 10^{-9} \,\mathrm{m}$
2 $1.29 \,\mathrm{kg\,m^{-3}}$
3 1.1 mol; 6.7×10^{23}; $2.1 \times 10^{25} \,\mathrm{m^{-3}}$
4 $2.5 \times 10^{-5} \,\mathrm{m^{3}}$
5 $8.7 \times 10^{-6} \,\mathrm{m^{3}}$

Section 4.2 (page 66)

1 212 kPa
2 1.28
3 $5.8 \times 10^{3} \,\mathrm{N}$

Now it's your turn (page 68)

$2.6 \times 10^{-10} \,\mathrm{m}$

Now it's your turn (page 72)

$1.04 \times 10^{-20} \,\mathrm{J}$

Now it's your turn (page 78)

1 (a) 1400 J gain   (b) 41 kJ loss
2 $4200 \,\mathrm{J\,kg^{-1}\,K^{-1}}$

Now it's your turn (page 79)

(a) (i) 16.5 kJ (ii) 113 kJ
(b) 6.8

Now it's your turn (page 80)

480 °C

Section 4.6 (page 81)

1 (a) 35°C

Exam-style Questions (page 81)

1 (a) (i) $2.69 \times 10^{25} \,\mathrm{m^{-3}}$ (ii) $3.3 \times 10^{-9} \,\mathrm{m}$
   (b) 4×10^{-4}
2 (a) 939 °C (1212 K)   (b) 239 kg
3 $2.4 \times 10^{5} \,\mathrm{Pa}$
5 (a) 16   (b) 1
   (c) 16 (in each part, the ratio is hydrogen : oxygen)

Chapter 5

Now it's your turn (page 85)

$2.3 \times 10^{-8} \,\mathrm{N}$

Now it's your turn (page 89)

(a) $5.0 \times 10^{4} \,\mathrm{V\,m^{-1}}$ (b) $2.4 \times 10^{-12} \,\mathrm{C}$

Now it's your turn (page 90)

$1.0 \times 10^{6} \,\mathrm{N\,C^{-1}}$

Now it's your turn (page 91)

26 V

Section 5.1 (page 92)

3 (a) $2.1 \times 10^{21} \,\mathrm{N\,C^{-1}}$ (b) 650 N

Now it's your turn (page 94)

1.0 mF

Now it's your turn (page 95)

0.203 J

Now it's your turn (page 98)

(a) 9.4×10^{-3} C
(b) (ii) 940 μF (iii) 10 V (iv) 4.7×10^{-3} C

Now it's your turn (page 101)

(a) 6.0×10^{-5} C (b) 2.7×10^{-5} C (c) 5.4 V

Section 5.2 (page 102)

1 (a) 1.1×10^{-10} F
   (b) (i) 1.3×10^{-9} C (ii) 2.4×10^{4} V m^{-1}
2 (a) 2.0 μF (b) 1.2 μF
3 (a) (i) 4.1×10^{-4} J (ii) 9.0×10^{-5} C
   (b) (i) 6.0×10^{-5} A (ii) 1.5 s (iii) 0.77 s

Exam-style Questions (page 103)

1 (a) 1.0×10^{4} V m^{-1} (c) 5.0×10^{-5} N
2 2.2×10^{39}; 2.2×10^{39}
3 (a) 2.8×10^{-2} N (c) 1.9×10^{-7} C

Chapter 6

Now it's your turn (page 106)

1 Diagram should be as in Figure 6.4 but with direction of field lines reversed.
2

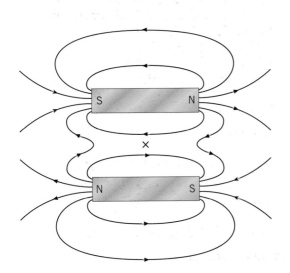

Now it's your turn (page 109)

1

2

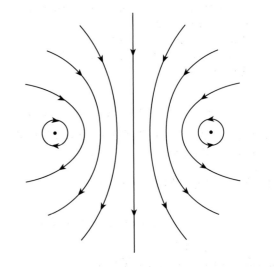

Now it's your turn (page 114)

1 3.4×10^{-7} N m^{-1}
2 (a) 2.8×10^{-2} N m^{-1} (b) 2.0×10^{-2} N m^{-1}

Now it's your turn (page 118)

The sphere moves downwards with an initial acceleration of 1.6 m s^{-2}.

Now it's your turn (page 120)

(a) 8.8×10^{6} m s^{-1} (b) 9.3 cm

Now it's your turn (page 122)

4.2×10^{6} m s^{-1}

Now it's your turn (page 124)

$m = qB^2r^2/2V$; without a velocity selector, the ions are unlikely to all have the same speed, as some may lose energy by scattering.

Now it's your turn (page 126)

1 (a) 5.0×10^{-14} N (b) 8.8 cm
2 Deflections in opposite directions because the particles are oppositely charged; electron has smaller radius of path because q/m is larger.

Section 6.2 (page 127)

1 (a) (i) east–west   (ii) 2.5×10^4 A
    (b) No, large current would melt the wire.
2 (a) (i) 1.5×10^{-4} T   (ii) 2.3×10^{-3} N m^{-1}
    (b) Forces not large enough to move copper wire.
3 (a) 0.15 T
    (b) uniform field, no field outside poles, wire normal to the field
4 (a) left to right using $r = mv/Bq$
    (b) Positive, with argument based on Fleming's left-hand rule.

Now it's your turn (page 135)

1 (a) (i) 1600 m^2 (ii) 0.064 Wb   (iii) 0.064 V
    (b) wing pointing west
2 86 mV
3 (i) 2.7 m^2 (ii) 0.41 Wb   (iii) 0.41 V

Now it's your turn (page 138)

1 (a) 45   (b) 56 mA
2 (a) 30   (b) 7500   (c) 450 mA

Section 6.3 (page 140)

1 (a) (i) 1.4 T   (ii) 4.0 mWb   (iii) 1.6 Wb
    (b) 150 V across switch
    (c) Large voltage induced which could be dangerous.
2 (a) 0   (b) 50 mV   (c) 29 mV
3 (a) 0.034   (b) 4100
    (c) Thicker wire reduces heating effect caused by the larger current.

Exam-style Questions (page 141)

1 (a) north–south
    (b) (i) east–west   (ii) 26 μT   (iii) 55°
    (c) 2.0×10^{-4} N m^{-1}
2 (a) (i) 4.5 mT   (ii) 1.1×10^{-4} Wb
    (b) 7.1 mV
3 $E = 0$ when current constant. Spikes in opposite directions at times t_{on} and t_{off}.

Chapter 7

Now it's your turn (page 144)

19, 40, 21

Sections 7.2–7.4 (page 153)

3 $^{a-4}_{b-2}$Y, $^{a-4}_{b-1}$Z, $^{0}_{0}$γ

Now it's your turn (page 156)

(a) 1/4   (b) 1/1024   (c) 3/4   (d) 15/16

Now it's your turn (page 159)

(a) 5.6×10^{12} Bq
(b) 2.5×10^{6} Bq

Now it's your turn (page 160)

(a) 3.3×10^{12} (b) 1.3×10^{12}

Now it's your turn (page 161)

1 (a) 1.6×10^{-4} s   (b) 1.7×10^{11} s
2 (a) 1.2×10^{19} s^{-1} (b) 4.6×10^{-2} h^{-1}

Sections 7.5–7.6 (page 162)

1 (b) 6 minutes
2 2.5×10^{9} Bq
3 (a) X: 3.5×10^{5} Bq; Y: 2.0×10^{5} Bq
4 1.2×10^{-10} kg
5 (a) 2.2 h   (b) 1.3×10^{21}
6 1300 years

Now it's your turn (page 164)

0.109906 u

Now it's your turn (page 165)

101 MeV

Now it's your turn (page 166)

7.3 MeV per nucleon

Sections 7.7–7.11 (page 172)

1 (a) 0.00857 u, 7.98 MeV, 2.66 MeV/nucleon
    (b) 0.687 u, 640 MeV, 6.60 MeV/nucleon
    (c) 1.79 u, 1660 MeV, 7.50 MeV/nucleon
2 (a) 236.132 u; 235.918 u
    (b) 0.214 u   (c) 208.5 MeV
    (d) 8.13×10^{10} J   (e) 74 g
3 (a) 4.96 MeV   (b) 14 MeV

Section 7.12 (page 176)

1 Ratio = 6.2

Now it's your turn (page 180)

1 baryon number = 1; charge = 0
2 baryon number = −1; charge = −e

Section 7.13 (page 181)

1 (a)(ii) neutron, proton

Exam-style Questions (page 182)

1 15.8 years
2 (a) $P = 131$; $Q = 54$
   (b) (i) 9.2×10^{20} **(ii)** 0.032 g
3 (a) $0.046\,h^{-1}$
   (b) $6000\,cm^3$
4 (a) 3.27 MeV
5 (c) $3.0 \times 10^5\,m\,s^{-1}$
6 (a) about 50 mm; $3300\,mm^{-1}$
7 (a) leptons = 4 (electrons);
       baryons = 9 (protons and neutrons)
   (b) X = neutrino

Chapter 8

Now it's your turn (page 187)

(b) $2.8 \times 10^{-11}\,m$

Now it's your turn (page 191)

1 23.8 cm
2 0.19

Now it's your turn (page 200)

(b) 9.4×10^{-3}

Section 8.3 (page 204)

2 (a) 0.004 I reflected at boundary
       0.994 I transmitted at boundary
   (b) 0.37 I reflected
       0.63 I transmitted

Exam-style Questions (page 213)

2 (c) 7.7×10^{-4}

Chapter 9

Now it's your turn (page 219)

65 years

Now it's your turn (page 223)

7.28 MeV

Now it's your turn (page 229)

$0.5c$

Exam-style Questions (page 237)

2 (b) $2.4 \times 10^{-18}\,s^{-1}$
3 (a) **(i)** $10^3\,K$ **(ii)** $10^{13}\,K$ **(iii)** $10^{27}\,K$
   (b) **(i)** $10^{-3}\,S$ **(ii)** $10^{-13}\,S$ **(iii)** $10^{-27}\,S$

Index